Advances in Anatomy
Embryology and Cell Biology

Vol. 141

Springer

Berlin
Heidelberg
New York
Barcelona
Budapest
Hong Kong
London
Milan
Paris
Santa Clara
Singapore
Tokyo

Magdalena Müller-Gerbl

The Subchondral Bone Plate

With 29 Figures and 16 Tables

 Springer

MAGDALENA MÜLLER-GERBL
Anatomische Anstalt I
Pettenkoferstraße 11
80336 München
Germany

ISBN 3-540-63673-0 Springer-Verlag Berlin Heidelberg New York

Library of Congress-Cataloging-in-Publication Data
Müller-Gerbl, Magdalena: The subchondral bone plate / Magdalena Müller-Gerbl. p. cm. – (Advances in anatomy, embrology, and cell biology; Vol. 141)
Includes bibliographical references and index.
ISBN 3-540-63673-0 (soft: alk. paper)
1. Bones, 2. Bone densitometry, 3. Joints, 4. Biominerlization, I. Title, II. Series. QL801.E67 Vol. 141

© Springer-Verlag Berlin Heidelberg 1998
Printed in Germany

Production: PRO-EDIT GmbH, D-69126 Heidelberg
SPIN: 10644953   27/3136-5 4 3 2 1 0 – Printed on acid-free paper

*Dedicated to my husband Michael
my son Maximilian
and my mentor Professor Dr. R. Putz*

Contents

1 Introduction (Review of Literature)

1.1
Preliminary Remarks

The relationship between mechanical stress and bone morphology has been one of the focal points of research on the nature of the skeleton for more than a century (Wolff 1892; Fick 1911; Roux 1912). More recently, Thompson (1942), Cowin (1986), Frost (1986) and Carter (1987) were able to demonstrate that the morphology of any bone, at any single instant in time, is the sequential expression of all the stresses to which the tissue has been subjected over a long period. It was Braus (1920) who first put into words the idea that morphology is a historical record of events and Carter et al. (1989) used the extremely telling expression, "loading history" (of the bone).

Most authors start off with the assumption that bone is an optimal structure, in the sense that the greater the stress on any zone, the higher its density and/or the thicker its cortex (Wolff 1892; Pauwels 1965; Kummer 1972; Cowin and Hegedus 1976; Hegedus and Cowin 1976b; Hayes and Snyder 1981; Firoozbakhsh and Cowin 1981; Carter 1984; Frost 1986; Fyhrie and Carter 1986; Rubin and Lanyon 1987, Carter et al. 1987). It follows, therefore, that the principle on which a bone is built would involve the least amount of material possible necessary to achieve an optimum of resistance (lightweight construction).

Pauwels (1960), who succeeded Wolff (1892), Fick (1911) and Roux (1912) in investigating the interchange between form and function of the connective and supporting tissues, went further and was able to show that other components of the locomotor apparatus (cartilage, ligaments and tendons as well as bone) are, by their very anatomical structure, optimally adapted to their function, that is, to the main long-term stress of daily activity. Pauwels (1960) summarized these results in his theory of "causal histogenesis" (kausale histogenese), thus meriting the honor of being the first to present a conclusive theory formalizing the interchange between function and morphology in the supporting tissue. This new way of looking at the subject strongly emphasized the manner in which biological tissue, as opposed to technological building materials, reacts to mechanical stress. There is a fundamental difference here, since technological building materials can only respond to the action of forces with passive deformation (elastic or plastic deformation), thereby offering a greater or lesser strength or finally breaking down altogether. Biological material, on the other hand, has the additional property of being able to produce its own active response to stress. In tissue, external forces create compressive, tensile or shearing stresses and strains. Depending on the extent of the local deformation, this will lead, after variable lapses of time, to an increase or decrease of the material concerned.

Tillmann (1978a), in continuing Pauwel's work, used the hip joint to develop the idea that there is a particular range of stress, within the joint itself, to which the tissues react by increasing or reducing in size, thus responding to the altered mechanical situation. Should the local tissue deformation exceed a certain upper limit (σ_u), the result is the destruction of tissue. If, however, the deformation is kept, for some time, below the lower limit (σ_l), the tissue undergoes involution. Examples which support this theory are not confined to the joints, but can also be seen in the effects of physical training, where either an increase in the bone mass and strength of ligaments and tendons occurs (Tipton et al. 1970; Nilsson and Westlin 1971; Chamay and Tschantz 1972, Jones et al. 1977; Woo et al. 1981, Martin et al. 1981) or signs of immobilization appear with an accompanying reduction in bone mass, muscular atrophy, decrease in cartilage thickness and fall in the load-bearing capacity of tendons and ligaments (Issekutz et al. 1966; Donaldson et al. 1970; Noyes 1977; Whedon 1984). This can be observed in everyday clinical practice.

By applying the theory of causal histogenesis, one can regard the preservation or breakdown of individual supporting tissues as an understandable reaction to a given mechanical situation. Furthermore, the fact that individual tissue types arise from undifferentiated embryonic tissue (mesenchyme) under the influence of specific mechanical stimuli can also be accounted for convincingly. Prendergast et al. (1996), using biphasic finite-element models, demonstrated that tissue-level mechanical stimuli could control the differentiation of interface fibrous tissues.

These basic principles can be adjusted to apply to the components of a joint, as confirmed by examining the distribution of thickness throughout the articular cartilage; variations in the thickness were first noted in 1897 by Werner. Holmdahl and Ingelmark (1948) reported a correlation between cartilage thickness and the stresses acting upon it, that is, the distribution of stress within the joint. This work was continued and extended, chiefly by Kummer and his team (Oberländer 1977; Kurrat 1977; Kurrat and Oberländer 1978; Oberländer and Kurrat 1979), to include a whole series of joints. They suggested that the hyaline cartilage is thicker at those places in the joint where the stress is greater. Bullough and his coworkers (1985) further demonstrated that topographical differences, both morphological and biochemical, are to be found within the cartilage covering of the surface of a joint.

In a study of all the major joints, we (Müller-Gerbl et al. 1987a, b) evaluated the thickness of both the total cartilage and its calcified zone at regularly distributed points throughout the whole joint surface. In addition, we succeeded in showing that the thickness of the calcified zone in each joint had practically the same distribution pattern as that of the total articular cartilage itself.

Other authors, including Bennett et al. (1942), Trueta (1968), Stougard (1974) and Lane and Bullough (1980), stated that they too had found differences in the thickness of the calcified layer, corresponding to places in the joint where the loading was strong or weak. Since this thickness of the calcified zone shows the same pattern of distribution as the total cartilage thickness (Müller-Gerbl et al. 1987a, b), it does appear that the distribution of thickness in the calcified zone is dependent upon mechanical factors.

In addition, there are several joints in the human body where a concurrence can be demonstrated between the surface distribution of the predominant stress and a corresponding quantitative distribution of the subchondral bone density, that is, of the bone layer directly underlying the cartilage. This also applies to the hip joint; the most important investigations on this come from Pauwels (1965), Molzberger (1973), Oberländer (1973) and Kummer (1968). Oberländer's investigations (1973) could further show a similar distribution pattern of the thickness of the cartilage covering the joint.

During our own studies on the knee joints of dissecting-room bodies, where we simultaneously measured both the thickness distribution of the total cartilage and the calcified zone as well as the density distribution of the subchondral bone in the same joint, we found that the individual comparisons showed no uniform type of distribution, such as that associated with the hip joint. There were, in spite of a general surface similarity, some marked individual differences in the six specimens which related to the position of the cartilage maxima in the femoral condyles and patellae. We interpreted these differences as a form of adaptation to the individually differing loading patterns, such as those found in genu valgum or varum.

As with the hip joint, however, there was agreement between the distribution of each of the three parameters within one individual. Our own comparison of the three parameters in the large joints of the lower limb disclosed a relationship between the stresses on the joint, the thickening of the cartilage and the mineralization of the underlying bone. This led us to conclude that the quantitative distribution of both subchondral bone density and cartilage thickness is a direct indication of local adaptation to the pressure transmitted through the joint.

1.2
General Mechanisms for the Regulation of the Morphology of Bone Tissue

Ever since the classical publication by Julius Wolff (1892), numerous investigations have been undertaken to confirm his *Law of the functional adaptation of bone*. However, as has already been stressed, the first really successful attempt to do this was made by Pauwels (1965). Kummer (1962, 1972, 1978) continued his work and interpreted the adaptation of the bone as a cybernetically controlled process. According to this, the remodelling of bone is determined by the degree of local stress, in such a way that the amount of bone is proportional to that stress and so "a body of equal strength" is produced. Kummer distinguished between a long-term mechanism (remodelling of the organic matrix) and a short-term mechanism (the deposition or resorption of calcium).

To account for the versatility of the adaptation, one need only presuppose a single property of the bone tissue; namely, that the relationship between the continuous bone deposition and resorption is dependent upon the local stresses acting on the bone. If this prerequisite is accepted, it must follow that, for a particular degree of stress (which Pauwels designated the "boundary stress σ_b"), deposition and resorption of bone tissue will remain in balance, so that the total amount of bone remains the same and the tissue is in equilibrium. Within certain physiological limits, the greater the stress, the greater the deposition and vice versa. This response is confirmed by observing the hypertrophy that accompanies activity and the atrophy that is produced by inactivity.

A similar hypothesis was developed by Carter (1984). He extended the observations of Pauwels and Kummer by demonstrating that immature bone reacts more strongly to mechanical influences (strains) than mature, adult bone. However, one question remains unanswered: what are the actual values of the boundary stress, i.e., the stress that limits the region of physiological response? In the meantime, there have been several quantitative studies of bone remodelling carried out, both in vivo (Lanyon et al. 1982; Goodship et al. 1979) and in vitro (Steinberg et al. 1974). Moreover, they were performed on animals and under oversimplified conditions of loading that can easily be understood and which exclude complex influences, so that a realistic situation is not adequately represented. For this reason, no data applicable to humans are available. There is no doubt whatever that other factors, such as the action of the endocrine system or metabolic changes, can affect the composition of bone. For local changes, however, effective, mechanical factors are decisive.

Arising from the fact that mechanical factors are responsible for the local deposition of bone, there is a further question which goes beyond the limits of this present work; namely, how is mechanical action on a tissue finally converted into information that the affected cells can recognize? A change in the mass of the bone can only be brought about by boosting or otherwise altering the production of matrix. How, in other words, does the cell "know" that it is more or less under strain and, thus, become able to react with an increase or decrease in matrix production? Here again, certain indications exist. Authors, such as Somjen et al. (1980), Alexander et al. (1984), Bindermann et al. (1984) and Skerry et al. (1987, 1988), have observed an increase in certain substances (e.g., prostaglandin E2, cAMP and cGMP) in cell cultures under conditions of increased stress and have expressed the opinion that these substances may be capable of transmitting information about the mechanical conditions prevailing. Other authors (Rodan 1981; Korenstein et al. 1983) have considered piezoelectric phenomena to be responsible for conveying this information. Ypey et al. (1991) regard the ionic canals of the cell membranes of osteoblasts and osteoclasts as sensory receptors that register and convey information about the mechanical situation and are, therefore, responsible for regulating osteocytic activity so that it may adapt itself to local stress. These authors consider it conceivable that changes in the stream passing through the membrane may alter the concentration of ions (K^+, Na^+, Ca^{2+}, Cl^-) within the cell and that these serve as intracellular signals, which regulate cellular activity in an adaptive response to the mechanical stimulus. Cowin et al. (1995) view the strain-generated potentials in bone tissue as part of a more complicated mechanism, in which the transmembrane potential and intracellular current through gap junctions are modulated by a fluid shear-stress-dependent intracellular biochemical response.

Since even short-term loading (measured in s) can produce an increase in matrix production, in a long-term response, the question arises of where this mechanical information is stored naturally. Skerry et al. (1988) has been able to show that a short period of dynamic stress alters the orientation of the proteoglycans in bone tissue; a change that slowly returns to the original condition within 48 h. This rearrangement may constitute a means of memorizing the result of the stress.

A hypothesis formulated by Mullender et al. (1994) says that bone contains mechanoreceptors, which implies that the regulation of bone mass is governed locally by sensor cells. According to Yates (1987), it is assumed that bone mass regulation occurs at a local level, which is typical for a self-organizational process. All these studies showed that bone and cartilage cells are sensitive to mechanical stimuli. A review of such studies is given by Burger and Veldhuijen (1993).

These constructional elements, which indicate the means of conveying mechanical information, can be summarized in the following manner, according to Stockwell (1987). As the final links in this chain of signals, cAMP and calcium are those which trigger the cellular response. As already indicated, mechanical stress produces a whole series of changes in the extracellular matrix that are partly mechanical and partly electrical in nature and, all of which, can call into play the enzyme systems associated with the membrane. Deformation of the tissue can raise the inflow of calcium by activating either the mechanosensitive calcium canals or the inositol system. At the same time, cAMP production can be increased by activation of the adenyl cyclase system. The calling into play of this system by the deformation of the cell membrane, with a corresponding release of strain energy, is also conceivable. The appearance of electric fields as a result of compression can also initiate an inflow of calcium by opening the stress-dependent ion canals. The influence of growth factors, interleukins and metallic proteinases has been disregarded.

It must, of course, be remembered that these studies were all carried out on bone cells and one must also take into account the fact that external influences in subchondral bone are of a different qualitative and quantitative nature from those in compact or cancellous bone. The pattern of subchondral mineralization cannot provide any clues about what is taking place at the cellular level, but it does allow one to judge which mechanical influences can lead to metabolic changes in the subchondral bone. Indeed, it makes it possible to interpret the mechanical conditions prevailing in small regions of the joint.

1.3
Morphology of the Subchondral Bone

1.3.1
Definition of the Term "Subchondral Bone"

Although there have been a great number of publications on the mechanical properties of normal and diseased articular cartilage, there is surprisingly little information available on the morphology and mechanical parameters of the layer of bone which lies immediately below it: the subchondral bone. Even the adjective "subchondral" is ambiguous, being used to mean several different things: (1) the calcified tissue immediately below the tidemark of the articular cartilage; (2) the thin cortical lamella directly underneath the radiologically discernible joint space; (3) the dense trabecular bone immediately adjacent to the cortical lamella; (4) a combination of 1, 2 and 3; and (5) the total subarticular tissue. Duncan et al. (1987) defined the "subchondral plate" as a zone which separates the articular cartilage from the marrow cavity and which normally consists of two layers: the calcified region of the articular cartilage and a layer of lamellar bone.

In this article, the term "subchondral zone" or "subchondral bone plate" is used to refer to the the bony lamella lying immediately beneath the calcified zone of the articular cartilage (Fig. 1). Depending upon the joint, this varies in thickness. The trabeculae arising from this bony lamella are referred to as "supporting trabeculae" and, together with the further bony components, are included under the term "subarticular".

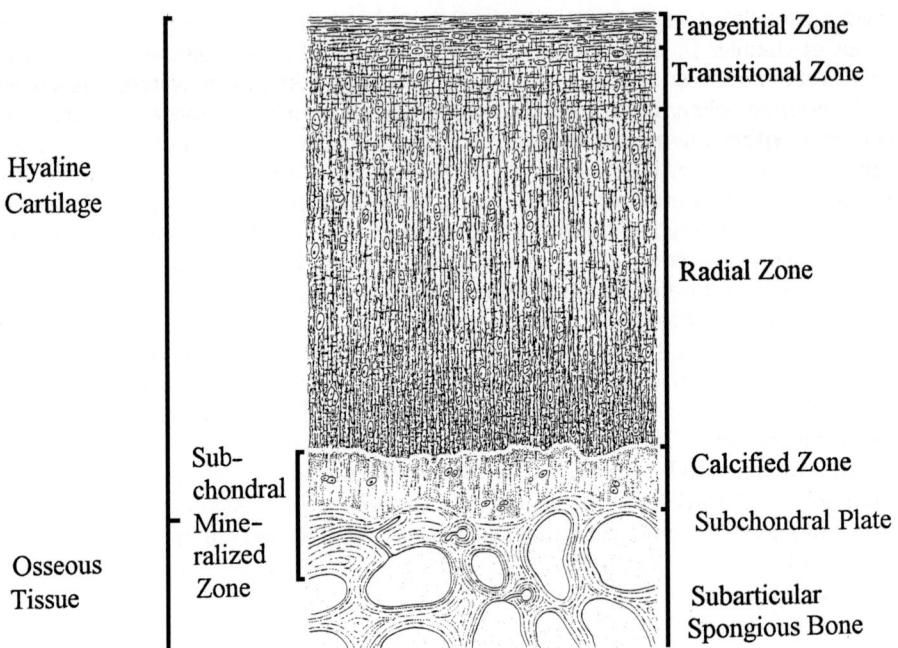

Fig. 1. Schematic drawing of the hyaline cartilage and the underlying subchondral bone

1.3.2
Structure of Normal Subchondral Bone

A number of studies have established that the anatomy of the subchondral region is highly variable. These variations include the contour of the tidemark and cement line, the number and type of perforations in the subchondral bone plate and its thickness, density and composition. Differences in the trabecular structure and mechanical properties of weight-bearing and nonweight-bearing areas have also been recognized.

1.3.2.1
Architecture

The subchondral bone plate consists of two mineralized layers which, together, form a single unit, separating the articular cartilage from the bone marrow. On the articular side, at the line of contact between the subchondral plate and the articular cartilage, there is a discrete band of mineralized cartilage which is more radio-dense than the adjacent bone. It has an affinity for numerous histological stains and appears as the "tidemark" in hematoxylin sections. From the tidemark, the calcified cartilage extends for a varying distance toward the marrow cavity, where it is remodelled and replaced by woven or lamellar bone, similar to the supporting trabeculae. These trabeculae, which are predominantly perpendicular to the joint surface, are themselves crossed at right angles by finer trabeculae. Very often, they are lamellae or sheets rather than actual trabeculae. (Duncan et al. 1985, Meachim and Allibone 1984).

When the dense cancellous bone directly underlying the calcified cartilage is seen in a section at right angles to the articular surface, it has the appearance of a solid mass of bone, fenestrated by numerous intercommunicating spaces. When, however, it is viewed in a section tangential to the articular surface, it is clear that the bone consists of plates which join together to enclose intervening spaces. The plates are relatively thick (0.2–0.4 mm) and the spaces relatively narrow (0.4–0.6 mm). Just below the articular surface, these spaces are roughly as long as they are broad, so that one is reminded of a honeycomb (Singh 1978). This appearance is, however, only seen for a few millimeters below the articular surface. Deeper in the bone, the spaces enlarge considerably and gradually tend to become elongated in a direction at right angles to the articular surface.

Observation of vertically cut surfaces with the scanning electron microscope (Inoue 1981) has shown that the subchondral plate under the osteochondral junction is composed of lamellated sheets of parallel collagen fibrils which continue into the lamellae of the bone trabeculae. Occasional vascular channels are connected to the osteochondral junction by small canals.

In the partially decalcified samples of subchondral bone, each trabecula was seen to consist of between 10 and 15 layers of lamellated sheets of collagen, 10–20 mm in thickness. A higher-powered view of the trabeculae revealed that each of the lamellae was composed of dense fibrils crossing each other among the lamellar sheets. In the almost completely decalcified samples, it could be seen that the intramedullary surface of the trabeculae consists of densely woven collagen fibrils, oriented obliquely to the trabeculae in layers, collected into bundles and interconnected to form a thick network. Some elliptical cavities are present among the network and probably correspond to osteocytes or canaliculi. In deproteinized samples, the SEM brought out the fine alignment of the mineral content on both the cut surface and the medullary surface of the subchondral trabeculae. The lamellar alignment of granular particles (probably hydroxyapatite) follows the orientation of the collagen fibrils seen in the decalcified samples. Holes of various sizes, which represent vascular canals and osteocyte lacunae, were found in some areas.

1.3.2.2
Vascularity of the Subchondral Bone Plate

The whole of the subchondral plate is invaded by hollow spaces which can be recognized both in histological sections and on contact radiographs and which provide a direct connection between the uncalcified cartilage and the marrow cavity of the spongiosa. These spaces can be allotted to groups according to their shape (Holmdahl et al. 1950; Inoue 1981; Green et al. 1970 and Milz and Putz 1994a). Where the subchondral plate is thin, they tend to be wide and resemble ampullae. In regions of greater thickness, however, channels can be seen. These are narrower and form a tree-like mesh. Occasionally, they involve the thicker subchondral-supporting trabeculae.

Green et al. (1970) observed these canals predominantly in young people still growing, but with increasing age their number began to fall away. The development of holes in older people is often regarded as the result of pathological processes associated with arthrosis (Green et al. 1970; Oettmeier et al. 1989). Green et al. (1970), therefore, came to the conclusion that the larger part of the subchondral mineralization zone in older adults represents devitalized tissue which is liable to induce resorption and invasion by blood vessels. In contrast, Milz and Putz (1994a) found these canals regularly and relatively frequently in older people, suggesting that a simple relationship with the processes of growth is unlikely. Furthermore, these authors found no indication that a large part of the subchondral mineralization zone in the adult consists of devitalized tissue.

As a result of SEM examination of the subchondral pressure plate of the human tibia, Duncan et al. (1987) were able to establish the presence of many perforations, a large number of which bring the articular cartilage into direct contact with the marrow cavity and, under the light microscope, many blood vessels with occasional erythrocytes could be recognized. The distribution of the perforations in the medial and lateral parts of the tibial plateau is not the same. On the medial side, the authors found most of the perforations in the dorsal region of the joint below the meniscus; whereas, on the lateral side, they were preferentially assembled at the center of the joint surface and not grouped together below the meniscus. Whereas Duncan et al. (1987) did not group the perforations in terms of the morphology, Milz and Putz (1994a), confirming the observations of Berry et al. (1986), found that these canal-like holes were preferentially grouped together in the subchondral plate of the tibial plateau at the central part of the joint, that is to say, in regions where the stress is greatest (Ahmed and Burke 1983).

The concentration of such hole formation in heavily stressed zones (Berry et al. 1986) suggests that the cartilage there, and certainly in the adjacent bone also, have a particularly rich blood supply. Through these canals, the blood vessels can reach the overlying articular cartilage directly. This situation was also observed in the human femoral head by Mezaros and Vizkelety (1986). Lane et al. (1977), who undertook a quantitative study of the vascularity and a qualitative study of the remodelling of the calcified cartilage and subchondral bone endplate of adult and humeral heads in terms of age, showed that the vessel density was 15%–25% greater in heavily loaded joints or in heavily stressed regions of the same joint. They also observed a significant difference in the percentage of actively remodelling vessels between the more and less stressed portions of the femoral and humeral heads. Furthermore, the curves representing femoral vascularity and remodelling lay above those for the humeral head. It is known that the magnitude of the load transmitted through the hip is, in general, greater than that acting on the shoulder joint. This suggests that vascularity and remodelling in the subchondral bone varies, not only with the stress within the joint, but also with the stress in different joints. Haynes and Woods 1975, working on recently amputated lower limbs, also found differences in the number and location of subchondral perforations: more being present in the larger joints.

All these data confirm that a medullary pathway exists in normal human joints. Mital and Millington (1971), investigating the human hip joint with a stereoscan electron microscope, found canals lined with cellular material connecting cancellous bone and subchondral plate with the basal layer of the articular cartilage. These authors claimed that their findings supported the argument for a subchondral nutritional route. Where subchondral canals exist, nutrients can reach the basal cartilage via these perforations, as easily as diffusion occurs between the superficial and synovial fluid;

but where they are absent, the cartilage is dependent upon the synovial fluid as its sole source of nutrition.

Haynes and Woods (1975) emphasized that variations in the subchondral vascular anatomy may be related to the surface area, cartilage thickness and loading stress.

In conclusion, the most probable function of these spaces is to provide nutrition for the subchondral plate and the deeper layers of the articular cartilage. The previously prevailing view that this cartilage cannot receive any nutrition from beneath the subchondral plate, therefore, needs to be revised. Possibly, the spaces provide passage for the movement of fluid when the joint is heavily loaded, this being accompanied by both electrophysical and hydromechanical phenomena.

1.3.2.3
Thickness of the Subchondral Bone Plate

The thickness of the subchondral plate varies both within and between joints. In most joints, where the bones have a concave and convex component, the dome-shaped convex subarticular bony structures are thinner and more uniformly shaped than those in the complementary component, where the center of the cavity is associated with a much thicker subchondral plate than found peripherally (Simkin et al. 1980, 1991; Dewire and Simkin 1996).

In the case of the tibial plateau, some modification of these general observations has to be made, since it is almost flat. The presence of the wedge-shaped meniscal cartilage, with a thickness increasing from 0 mm at the center to between 5 mm and 7 mm at the periphery, produces a functional concavity which articulates with the condylar end of the femur. At the center of the contact area of each plateau, the subchondral bone is thicker than at the periphery (Duncan et al. 1985).

Clark and Huber (1990), who also investigated the structure of the subchondral plate in the tibial plateau, confirmed that it is narrow peripherally and thick near the center. Milz and Putz (1994b), who measured the thickness of the subchondral plate at about 600 points per tibial plateau, found a regular regional thickness distribution, decreasing outwards concentrically. The same is true of the 7 to 12-fold increase from the periphery to the center, which is more than the moderate difference reported by Duncan et al. (1987). For the first time, maps of the collective patterns of the subchondral plate (Milz and Putz 1994a, b) appeared in the literature. Similar maps have also been made of the articular surface of the patella (Milz et al. 1995) and the trochlear notch of the ulna (Milz et al. 1997).

1.3.2.4
Density Distribution in the Subchondral Plate

Regional differences in the density distribution (mineralization) of the subchondral bone can also be recognized and greater density is regularly found in the more heavily loaded regions of the joint surface.

The density distribution in the tibial plateau was investigated by Noble and Alexander (1985). Their data, as well as the findings reported by Schünke et al. (1987), who found greater subchondral bone density in the medial than in the lateral part of the plateau, were essentially in agreement with the thickness distribution of the subchondral plate reported by Milz and Putz(1994b). Similar results were also reported by Odgaard et al. (1989), who found a more pronounced density maximum dorsally displaced in the medial articular surface, whereas that of the lateral surface was central and less prominent.

In contrast to the situation in the femorotibial joint, the zones of highest density were constantly found in the lateral facet of the femoropatellar joint (Pedley and Meachim 1979; Eckstein et al. 1992).

Oberländer (1973), who examined the density distribution in the human acetabulum, was able to demonstrate two different age-dependent patterns. There is either a monocentric pattern with a central maximum or – in one specimen – a bicentric maximum with dorsal and ventral maxima and a significantly lower density between them.

An investigation by Möllers et al. (1986) on the wrist joint revealed that there are always two maxima on the radial articular surface, each of which is centrally located.

All these studies have confirmed that every joint surface has regular reproducible patterns of density distribution and it can be suggested that these are correlated with the mechanical situation in the joint and reflect the long-term stress acting there.

A comparison between the various studies of the thickness and density distribution and the distribution of the vascular density of the subchondral plate (particularly in the tibial plateau, where they are investigated most often) shows that these findings are in good agreement with each other. At places within the joint where the stress is assumed to be greatest, the density is higher, the thickness greater and the vascularization more strongly developed. A major disadvantage of these earlier studies is that they were all involved with postmortem material.

1.3.3
Structure of Pathologically Altered Subchondral Bone

1.3.3.1
In Osteoarthritis (OA)

An equally large number of investigations into the alterations taking place in the subchondral bone, in the presence of osteoarthritic change, have been carried out which involve the architecture of the plate as well as the thickness and density distributions, the mechanical conditions and the vascularization.

OA is a very common, slowly progressive, degenerative disorder of synovial joints affecting both the articular cartilage and the underlying subchondral bone. Although most investigators have focused on the articular cartilage to identify the earliest changes in OA (Bland 1983), a few have concentrated their efforts on studying subchondral bone remodelling, since this could be involved in the primary pathological process (Radin et al. 1970). These results are sometimes contradictory.

In 1969, Sokoloff studied the pathology of OA in human hip joints and found that cartilage fibrillation could not be dissociated from the bony changes, even in the earliest stages of the disease. Mayor and Moskowitz (1974) carried out an autoradiographical study on experimentally induced OA in rabbits, in which the earliest evidence of cell replication and degenerative biochemical changes were observed simultaneously in the articular cartilage and subchondral bone.

Grynpas et al. (1991) assessed the degree of mineralization by density fractionation and chemical analysis, and compared the histomorphometrical findings in osteoarthritic human femoral heads with age-matched and young controls. They found in both the weight-bearing and nonweight-bearing regions of the subchondral bone of the osteoarthritic specimens less mineralization than in the age-matched and young controls. Histomorphometric analysis showed that the subchondral bone thickness, all osteoid parameters and the eroded surfaces were increased in the osteoarthritic specimens in comparison with the controls. Mineralization in the deep cancellous core bone increases with normal aging but undergoes less change in OA. The cancellous core showed that the osteoid parameters, but not the bone volume, increased more in the OA specimens than in the controls. From this, one may conclude that OA is associated with a thickening of subchondral bone with an abnormally low mineralization pattern.

A study by Inoue (1981), dealing with alterations in the collagen framework of osteoarthritic cartilage and subchondral bone, disclosed the vertical arrangement of the collagen in all layers. The fibrils were organized into large bundles which were sometimes separated down as far as the calcified layer and which were extensively frayed in the superficial layer. In the deeper layers, some of the fibrils mingled with the subchondral bone plate to produce an irregular or indefinite tidemark and osteochondral junction. Granular particles (0.5 μm) of hydroxyapatite were seen to be deposited throughout the deep layer, extending from the calcified layer around the tidemark. The collagen in the subchondral bone was irregularly laid down, unevenly distributed on the surface of the trabeculae. The absorption phase was represented by many lacunal cavities with newly formed lamellae. It is possible that this irregular architecture of the subchondral bone may represent an early stage of arthritic change, as suggested by Lereim and Goldie (1975).

The finding by Christensen et al. (1982) of large quantities of woven bone at the overloaded condyle is suggestive of an early "productive" phase in the development of osteoarthrosis. According to the SEM investigations of Clark and Huber (1990), normal subchondral bone is mainly constructed of appositional layers continuous with trabeculae; Haversian canals were seldom seen. However, where the subchondral bone was exceptionally thick beneath degenerated cartilage, it was composed primarily of an irregular network of osteons.

Other studies of osteoarthritic hips (Reimann et al. 1977: Cameron and Fornasier 1979; Reimann and Christensen 1979) have reported that bony changes (including trabecular bone volume, mineral ash and alkaline and acid phosphatase activity) were found to be greater in the weight-bearing region. In the osteophytes, the activity of both enzymes was rather high. By comparing the enzyme activity (alkaline and acid phosphatase, used as marker enzymes for bone deposition and resorption, respectively) by means of histological/histochemical grading of the cartilage, Reimann and Christensen (1979) found a significant direct correlation. This was assumed to be an expression of the increased activity of both enzymes in osteoarthritic subchondral bone, where it presumably reflects an increase in both the deposition and resorption of bone. In 67% of the osteoarthritic femoral heads studied histologically by Batra and Charnley (1969), osteoid was found exclusively in the pressure zones. Christensen et al. (1982) used photon absorptiometry to demonstrate that osteoarthritic subchondral bone in loaded tibial condyles has an increased mineral content and double the percentage volume of trabecular bone.

In patients with OA, Havdrup et al. (1976) found a marked tendency for the subchondral bone of the tibial plateau to undergo sclerotic change. This took the form of thickening and confluence of the trabeculae into massive blocks of bone containing single vascular canals. When the bone was completely denuded, the surface layer exhibited necrotic osteocytes. The severity of the sclerosis varied with the damage to the overlying cartilage, the underlying trabeculae being of normal appearance when the cartilage was relatively intact. The thinner the cartilage, the more massive the clearly focal sclerosis was. These sclerotic changes varied in depth from 500–3000 μm, but sclerotic bone was often in direct contact with cancellous bone of normal appearance, indicating that the architecture of the subchondral bone trabeculae was also irregular.

Microscopic computed axial tomography was used by Layton et al. (1988) to examine the trabecular architecture of the subchondral bone of the femoral head in a guinea-pig model of OA. They found that the trabeculae were thicker and closer together, resulting in a highly significant increase in the bone fraction (the proportion of bone to marrow present in a given volume). This suggests that remodelling of the trabeculae may be an early event in the development of OA. The same technique was also used to study the subchondral plate and trabecular bone in a canine experimental model of OA induced by cutting the anterior cruciate ligament (Dedrick et al. 1990). The subchondral plate was found to be slightly thinner in the operated knee, although this difference was not statistically significant. The bone fraction was significantly decreased relative to the contralateral knee at 13 weeks, but no significant difference was apparent at 72 weeks. These changes are consistent with a slight loss of bone due to decreased weight bearing by the operated limb. Nevertheless, subchondral sclerosis was seen on straight radiographs at 72 weeks.

Kamibayashi et al. (1995) demonstrated that, in tibial plateaus of patients collected during hemiarthroplastic and total knee arthroplastic operations, in comparison with age-matched controls, there is a significant increase in the bone volume fraction and trabecular thickness in OA. Overall trabecular orientation in the osteoarthritic group was more vertical or perpendicular to the articular surface than in the control group, especially in those trabeculae in the cancellous bone layer closest to the articular surface. These structural changes may ultimately lead to further joint degeneration as the bone stiffness gradients depart significantly from normality, as already suggested by Radin et al. 1970. Although osteoarthritic bone adapts functionally to the altered mechanical-loading conditions, this adaptation may ultimately hasten failure of the joint.

Besides changes in architecture, density, thickness and mechanical properties, alterations in the vasculature have been described in OA. Harrison et al. (1953) stated that the number of subchondral vessels is increased and that this is a response to the disease rather than its cause. According to Harrison, excessive loading on the hip joint "wears out" the articular cartilage and accelerates its degeneration by eliciting invasion by the juxtachondral blood vessels, thus "weakening" the subchondral bone. Increased vascularity of the subchondral bone is typical of joints that develop OA, particularly the hip and knee joints (Harrison 1953; Mankin 1974; Sokoloff 1974; Stevens 1970). These findings suggest that OA is a natural sequel to vascular invasion and increased remodelling of the subchondral region, which may themselves constitute an attempt to reestablish a proper distribution of the load. On the other hand, increased subchondral bone density may concentrate stress in the overlying cartilage, thus rendering the reparative action of the joint ineffective or even damaging the overlying cartilage.

Ogata et al. (1977) produced varus stress in nonimmobilized rabbit knees in order to bring about a gradually progressive unicompartmental OA without opening the joint. After 68 weeks, they observed increased vascularization of the subchondral bone in areas adjacent to severely damaged cartilage. There was, however, no noticeable thickening of the bony trabeculae.

The results of a morphometric study of tibial bone by Shimizu et al. (1993) indicated that there is a correlation between the degree of macroscopic or microscopic degeneration of the joint cartilage and the bone volume and rate of formation in subchondral bone. A slightly increased rate of bone resorption was suggested only in the external and intermediate areas; it was extremely low in the lateral condyle. Throughout the regions observed, the ratio of bone volume was greatest in the superficial layer, gradually decreasing as the depth increased. Bone formation was most active in the superficial layer of the lateral condyle, whereas in the medial condyle it was at its height at a depth of between 1000 and 1500 µm. The rate of bone resorption did not vary with depth. Progressive degeneration of the joint cartilage was associated with remodelling of the underlying subchondral bone.

1.3.3.2
In Chondromalacia Patellae

The apparent similarity of the macroscopic and microscopic cartilaginous changes in the patella in chondromalacia patellae (CP) to those in joints affected by OA has frequently led to the suggestion that these conditions are etiologically related (see, amongst others, Gardner 1965). However, CP most commonly occurs between the ages of 16 and 30 years, whereas the incidence of OA increases with advancing age. Moreover, radiographical studies of affected patellae (Darracott and Vernon-Roberts 1971) have revealed that osteoporosis is always present, in contrast to the sclerosis which would be expected in OA. In a macroscopic and microscopic follow-up study, which included the use of radioisotope bone scans, Darracott and Vernon-Roberts (1971), working with autopsy material, observed hyperplasia of the chondrocytes with vascularization and advancing ossification into the deep zone of the articular cartilage, new bone formation and thinning of the subchondral osseous plate, focally or diffusely severe osteoporosis of the trabecular bone and, in some cases, superficial fibrillation of the cartilage and discreet nodular aggregates of woven bone on osteoporotic trabeculae. Radioisotope bone scans confirmed the presence of disordered calcium metabolism in this condition. All these findings suggest that CP is an etiologically different disorder, which is not due to OA or primary cartilaginous degeneration.

1.3.4
Mechanical Properties of the Subchondral Bone Plate in Normal and Pathological Conditions

Many authors have been concerned with the question of the correlation between the histology and mechanical properties of bone (among others, Aszenzi and Bonucci 1964; Evans and Bang 1967; Carter and Spengler 1978; Martin and Ishida 1989). In bone, many microstructural variables, or variables affecting the composition are relevant, including density, porosity, mineral or calcium content, anisotropy and the orientation of the collagen fibers. Attempts have been made to find a linear relationship between one of the above parameters of the material and its mechanical characteristics, such as strength, for instance. However, so far it is clear that there is no simple relationship between, for example, the strength of bone and its mineral content. Martin and Ishida (1989) were the first to successfully determine all the variables simultaneously and to correlate them with the strength of bone.

According to Wolff's law, the alignment of the trabeculae in cancellous bone follows the line of maximum principal compressive stress (Wolff 1892). Pugh et al. (1973a) found that the trabecular contiguity (a measure of the spatial arrangement and density of the trabeculae) contributes to stiffness, thereby reflecting the intensity of stress in the bone imposed during weight bearing. Radiographical and other densitometric analyses may be used to map cancellous bone density and, thus, define likely paths of load transmission. A more direct measure, however, may be obtained by examining the distribution of strength. Stiffness and compressive strength are two of the parameters most commonly used to describe the physical properties of bone. Compressive strength may be defined as the force at which the bone either starts to yield or falls under compressive loading. The "ultimate load" is defined as the force necessary to cause collapse at the point of insertion of the penetrator. Dividing the ultimate load by the original cross-sectional area of the penetrator elicits the ultimate strength. The elastic modulus of bone (ratio of stress to strain) is a measure of the elastic deformation before its collapse under loading. The ultimate strength of cancellous bone has a significant linear correlation with the elastic modulus (Goldstein et al. 1983; Linde et al. 1986; Harada et al. 1988). The "hardness" of bone is its property of withstanding the impact of a penetrating agent.

1.3.4.1
Strength of Subchondral Bone

Lereim et al. (1974) noted that the strength of a normal tibial condyle increases with age and that OA and rheumatoid arthritis (RA) render it weaker than normal. Bone strength has also been shown to vary from region to region of the transverse plane through the tibia (Behrens et al. 1974; Johnson et al. 1980; Lereim et al. 1974) and with the distance from the subchondral surface (Goldstein et al. 1983; Hvid and Hanson 1985). Hvid et al. (1985) found a wide variation in the maximal value between individuals, but the patterns obtained were remarkably uniform. In all but one knee, the medial condyle showed the highest peak value with a mean medial-to-lateral peak ratio of 1.7. In the medial condyle, the area of greatest strength was relatively large, with central and anterior peak values. The lateral condyle showed a more restricted, posteriorly localized area of greatest strength. Beneath the menisci, the bone strength decreased toward the margins of the condyles, and it also decreased to low values at the intercondylar region. A further study of the strength of the subchondral bone in the osteoarthritic knee (Hvid et al. 1986), where the strength values were obtained as a function of the location on the resection surface, revealed that in cases of genu varum the bone strength is greatest at the medial margin of the medial condyle. The patterns probably reflect the overall load distribution in the malaligned knee.

Gait studies, although dependent on a number of assumptions, have demonstrated the normal preponderance of loading in the medial compartment (Harrington 1983) and this is augmented by the varus deformity, but not always reversed in the valgus position. In view of the highly abnormal topographical distribution of bone strength following abnormal weight bearing at the knee, it is interesting to note that the remodelling activity of the subchondral bone is reduced in the osteoarthritic joint (Havdrup et al. 1976; Christensen et al. 1982). The finding of large quantities of woven bone in the loaded condyle (Christensen et al. 1982) is, however, suggestive of an early "productive" phase in the development of arthrosis and is possibly related to the presence of high strength values at the margin of the loaded condyle in genu varum and in some cases of genu valgum.

Harada et al. (1988) confirmed, in their study, that the medial condyle of the tibial plateau is much stronger than the lateral condyle and that, in both regions, the strength decreases rapidly with the distance from the surface, especially over the first 5 mm. In both condyles, the mean bone strength is greater in men than in women. The spatial distribution of the strength across the planes of the tibia seems to be consistent with the expected pattern of the load distribution and with contiguity of the trabecular bone.

Meanwhile, it is generally accepted that the medial condyle is stronger than the lateral. Based on these strength measurements, the proximal tibia may be mapped into strong weight-bearing areas (medial and lateral condyles), a weak central zone and paralateral and paramedial zones of intermediate strength. Photoradiographs of transverse sections from the tibia have shown that the areas of maximum strength correspond to regions where the trabecular bone density is greatest, providing, in effect, an image of the area mapped in terms of the strength. This confirms the association between trabecular contiguity and bone strength in relation to the concentration of strength within the bone (Pugh et al. 1973a). Functionally, the softer central region (intercondylar area) may serve to uncouple the medial and lateral condyles in load bearing and, thus, increase the tolerance for uneven weight distribution across the tibial plateau in normal knees (Andriacchi et al. 1986; Johnson et al. 1980; Kettelkamp and Chao 1972). Hvid (1988), investigating the axial strength of the trabecular bone of the tibial and femoral condyles in 150 consecutive total knee arthroplasties, found the tibial bone strength to be less in RA than in OA. Bone strength values at the tibia and femur varied significantly with the topographical position and the depth beneath the resection surface. In the medial tibial condyle, the bone was strongest centrally in OA, whereas in RA it was strongest posteriorly. In the lateral condyle, the strength was greatest posteriorly in both groups. In the femur the variation in bone strength was different. The highest values were obtained from the posterior aspects of the condyles, while the central areas were relatively weak. Paired comparison showed that the femoral bone is weaker than the tibial and that the strength tends to increase with increasing depth.

Behrens et al. (1974) suggested that there is a correlation of bone strength not only with the magnitude, but also with the duration and frequency of the normal stresses. It could be said that greater forces acting over a larger area of contact directly influence cancellous bone strength. In a joint with a varus or valgus deformity, the obvious difference in bone strength between the loaded and unloaded condyles also shows that cancellous bone responds to stress by remodelling. The nature of this functional adaptation is clearly illustrated by the radical alteration of both the amount of bone present and the trabecular pattern.

A number of factors could have a direct influence on the compressive strength. These include the bulk specimen density; the mineral content and material density and fractional area of the bone; the trabecular pattern; and possibly other microstructural parameters, such as the ratio of amorphous to crystalline calcium phosphate and the mineral to nonmineral bonding. From the studies reported here, it appears that specimen density and mineral content are important, but that the bone material density, although variable, does not correlate with its strength. A significant parameter is evidently the arrangement of the trabecular pattern, as revealed by both the different morphology and the 10:1 ratio found in fractional areas. The significance of the pattern distribution was pointed out by Pugh et al. (1973b). It seems that although variations in strength depend on the quantity of bone present, it is the reorganization of the trabecular pattern itself which is probably the single most important factor. This reorganization is likely to be evident in diseased joints (e.g., genu valgum or genu varum). It may well be that the strength adjusts to a certain level above the stress imposed so that it may remodel as necessary (Chamay and Tschantz 1972). A simultaneous investigation of the distribution of bone strength and cartilage thickness in the knee joint (including the patella and femoral and tibial condyles) showed that areas of thick cartilage correspond to areas of high subchondral bone strength in knees with no degenerative changes (Inaba 1996).

1.3.4.2
Stiffness of the Subchondral Bone Plate

Whereas several investigators have reported or predicted varying elastic moduli for cortical bone specimens of different sizes and conflicting data have been reported for the trabecular tissue, little information is available on the subchondral bone. Murray et al. (1984) calculated a Young's modulus of 3.06 (0.49 GPa for specimens from a human tibia. Brown and Vrahas (1984) tested specimens from human femoral heads and calculated moduli ranging from 0.82 (0.17 to 1.37 (0.41 GPa, and Choi et al. (1990) tested human subchondral bone beams from the proximal end of the tibia and calculated a modulus of 1.15 (0.37 GPa. Mente and Lewis (1994), who measured the elastic moduli of calcified cartilage and subchondral bone tissue experimentally for bovine patellae and the distal part of bovine femurs, obtained an elastic modulus for the subchondral bone calculated from "pure" subchondral bone beams of 2.3 (1.5 GPa (3.9 (1.5 GPa for the femur and 1.6 (0.7 GPa for the patella). The modulus for the calcified cartilage was more than one order of magnitude lower (0.32 (0.25 GPa) than the modulus for the underlying subchondral bone. This supports the suggestion that the zone of calcified cartilage forms a transitional zone of intermediate stiffness between the articular cartilage and the subchondral bone.

Choi et al (1990) determined the elastic moduli of human subchondral, trabecular and cortical tissue from the proximal end of the tibia, using three-point bending tests at a microstructural level. The mean modulus of subchondral specimens was significantly lower (1.15 GPa) than those of trabecular and cortical specimens (4.59 GPa and 5.44 GPa, respectively), in spite of the high mineral density. Earlier testing of relatively large subchondral specimens showed moduli ranging from approximately 1 to 3 GPa (Murray et al. 1984; Brown and Vrahas 1984). The microstructural characteristics, together with the presence of calcified cartilage in the subchondral specimens, might be responsible for the low modulus of the subchondral bone. According to the recent observations of Duncan et al. (1987) and Milz and Putz (1994b), the subchondral plate contains numerous perforations of various sizes which are not seen in trabecular and cortical bone. While the physiological function of the perforations (see Sect. 1.3.2.2) is not entirely understood, their mechanical effects are quite obvious. These microstructural defects will reduce the ability of subchondral bone to resist deformation, thus resulting in the low modulus. Furthermore, Choi et al. (1990) found a significant correlation between modulus and specimen size, whereby the surface area to volume ratio proved to be a key variable to explain the size dependency.

1.3.4.3
Energy Absorption Capacity of the Subchondral Bone Plate

In 1970, Radin and Paul were the first investigators to propose that subchondral bone stiffening caused by healing microfractures might be the primary pathogenic event in the development of OA. They carried out biomechanical tests on subchondral bone plugs removed at autopsy from the weight-bearing portion of the medial femoral condyle in 43 male patients with varying degrees of OA and from normal controls. There was a significant decrease in the energy-absorbing capacity of the subchondral bone in mild OA compared with the normal energy absorption in moderate and severe cases and in the controls. It was suggested that, in mild cases of OA, the articular cartilage is exposed to increased forces and eventually to wear because of the decreased energy-attenuating properties of the underlying subchondral bone. Further support for this mechanism was provided by Ewald et al. (1982) when they replaced the subchondral bone in living dogs with methacrylate cement which was 2.6 times stiffer than normal cancellous bone. After 2 years of normal ambulation, there was a uniform loss of the articular cartilage in the weight-bearing regions of all femoral heads.

Hayes and Carter (1976) tested cylindrical specimens of subchondral trabecular bone at uniaxial compressive strain levels of 75% to study energy absorption during collapse of the pores. The yield strength and energy absorption capacity were found to increase linearly with the apparent density of the specimen. Microstructural analysis of the deformed specimens confirmed that the mechanism for energy absorption was primarily the buckling and fracture of the trabeculae. These results suggest that, during fracture, the collapse of trabecular bone (together with the consequent absorption of energy) serves to attenuate stresses transmitted to the skeleton and, thus, protects vital structures such as the brain.

1.3.4.4
Hardness of the Subchondral Bone Plate

The hardness of bone is defined as its property of withstanding the impact of a penetrating agent. According to Currey (1970), this varies with the strength, the modulus of elasticity and the plastic flow that bone can undergo. The hardness of subchondral bone can be determined by an indentation method described by Brinell (1900). Lereim et al. (1974) demonstrated that the hardness of subchondral bone measured in the medial tibial plateau was significantly lower in the presence of OA and RA than in normal bone. They also noted that this difference in hardness was associated with destructive phenomena of various kinds, such as the invasion of granulation tissue, osteoporosis and signs of remodelling.

Björkström and Goldie (1982) analyzed the hardness of subchondral bone in the normal patella and in OA and in chondromalacia, which is regarded as a probable precursor of OA (see Sect. 1.3.3.2). They found the lowest hardness in normal bone. The mean hardness value beneath degenerated cartilage differed only slightly from that of the normal material, but the degree of variation was greater. The hardness of bone in an area of chondromalacia was less than in bone surrounded by normal cartilage. The mean hardness value in bone beneath areas of normal cartilage in specimens with chondromalacia was greater than that of the normal material. There was a tendency towards a relationship between variation in hardness values and bone structure. Thus, greater values in hardness were most often found to be associated with a confluence of the trabeculae to form a relatively dense plate of bone, with an increase of trabecular thickness and where thick transverse bars had joined the trabeculae together to produce a heavy mesh and with large irregular aggregates of bone in or near the subchondral plate. Lower hardness values were most often associated with either a thin subchondral plate, a decreased density of the bone plate or a less regular arrangement of the cancellous bone trabeculae, with thin trabeculae or with trabeculae oriented obliquely to the joint surface.

1.3.5
Correlation Between Structural Parameters and Mechanical Properties

To what extent the various mechanical properties are dependent upon the parameters of the material is a question of particular interest. Is there a simple linear relationship between the individual parameters or should a complex interdependence between the various factors be assumed?

The findings of Klawitter and Weinstein (1975) have indicated that the mechanical properties of subchondral cancellous bone are dependent on the volume fraction of bone, the mean cancellous strut thickness and the spacing.

Lereim and Goldie (1975) attributed hardness to the mineralization of the trabeculae, which increases with aging and is closely correlated with the trabecular architecture. The arrangement of the collagen and mineral substances also determines the hardness and elasticity of subchondral bone.

Evans (1969) suggested that the percentage of the fracture area formed by osteons and their fragments and by interstitial lamellae, regardless of the orientation of their collagen fibres, influences the strength and mechanical properties of bone.

According to Hvid (1988), the apparent density i.e., the defatted, hydrated tissue weight divided by the bulk volume (bone and empty spaces) of a specimen and the structural organization of the tissue are fundamentally important for the mechanical properties of trabecular bone (Behrens et al. 1974; Carter et al. 1980; Carter and Hayes 1976, 1977; Ducheyne et al. 1977; Favenesi et al. 1984; Fyhrie and Carter 1986; Galante et al. 1970; Hvid and Hanson 1985: Jasty et al. 1985a, b; Kaplan et al. 1985; Pugh et al. 1973b; Schoenfeld et al. 1974; Stone et al. 1983). In addition to this, the material properties of the solid phase affect the material properties of the structure (Carter and Hayes 1977). The compressive strength of cancellous bone has been shown to correlate closely with the apparent density (Galante et al. 1970; Carter and Hayess 1976) and with the mineral concentration (Weaver 1966; Galante et al. 1970). Using bovine and human trabecular bone samples and including results from tests on human and bovine cortical bone in their regression analysis, Carter and Hayes (1976, 1977) were able to show that the compressive strength is proportional to the square and the modulus of elasticity to the cube of the apparent density. Other workers, using trabecular samples from a single species, have found that a linear function predicts mechanical strength from density measurements equally well (Ducheyne et al. 1977; Galante et al. 1970; Hvid and Hanson 1985; Schoenfeld et al. 1974). According to Hvid (1988), the fundamental fact is that strength and stiffness are empirically closely related to the apparent density and that, in principle, this allows a nondestructive and noninvasive estimation of the material properties to be made from the density measurements.

The real density of trabecular bone, i.e., the defatted, hydrated tissue weight divided by the bone volume (excluding empty spaces) does not correlate positively with the compressive strength (Behrens et al. 1974; Galante et al. 1970). The concentration of the bone mineral (ash in mg divided by the apparent density volume) is, however, positively correlated with the compressive strength and stiffness (Behrens et al. 1974; Bentzen et al. 1987; Galante et al. 1970; Hvid and Hanson 1985), whether it is measured directly or indirectly by beam attenuation. Bentzen et al. 1987, who compared the Hounsfield values obtained from trabecular bone in the proximal end of the tibia by quantitative computed tomography (QCT), with mechanical data obtained from compression tests and penetration strength measurements, found close correlation between the relative linear attenuation coefficient (determined by CT Hounsfield unit = HU) and the ultimate strength ($r = 0.84$), yield strength ($r = 0.85$), elastic modulus ($r = 0.78$), ultimate energy absorption ($r = 0.83$), yield energy absorption ($r = 0.81$) and the penetration strength ($r = 0.82$). They concluded from this that these correlations are sufficient to make QCT a valuable noninvasive method for evaluating the spatial distribution of the properties of bone.

Hvid et al. (1989) explored the relationship of QCT to the apparent density and the ash-apparent density of tibial trabecular bone specimens and related these parameters to the results of destructive and nondestructive axial compression tests. Evaluation of the relationship of apparent density to Young's modulus and ultimate strength suggested that a power–law regression model is preferable to a linear model, although the linear prediction of mechanical properties is not significantly less accurate. Furthermore, they were able to show that with different energy values the results yielded closely related results ($r = 1.00$). The prediction of physically measured densities or mechanical properties is, therefore, not improved by using more than one scanning energy value. QCT and physically measured densities are intimately related (QCT at 140 kVp to apparent density using linear regression, $r = 0.94$ and to apparent ash density, $r = 0.95$) and do not differ significantly in their ability to predict the mechanical properties, thus favoring the simpler QCT for routine work.

Comparison of the distribution patterns of subchondral density, subchondral mineralization, vascularization density and the various mechanical parameters (such as strength, stiffness and hardness), obtained from different studies of the tibial plateau, reveals a striking agreement. It can also be assumed that the local expression of all these parameters depends upon the distribution of stress acting on the articular surface.

1.4
Function of the Subchondral Bone Plate

The subchondral plate is an integral and dynamic component of the joint and, so far as is known, one of its functions is to provide support for the overlying articular cartilage (Duncan et al. 1987). Its contribution to impulsive load attenuation (Pugh et al. 1974; Radin et al. 1970) and to endoprosthesis support (Pedersen et al. 1982) is well documented. According to Radin and coworkers (1970) and Pugh et al. (1973b), the subchondral bone absorbs most of the mechanical force transmitted by diarthrodial joints. Inoue (1981) claimed that the architecture of subchondral bone is adapted to its shock absorbing function during mechanical stress. Cartilage and bone act in concert by performing a mechanical function, the former as a bearing and the latter as a structural girder and shock absorber (Layton et al. 1988). It is, therefore, reasonable to consider OA not only as a disease of cartilage. Each of these anatomically closely related tissues is affected by any alteration in the mechanical properties of the other.

Because of its relatively greater stiffness and strength in comparison with the overlying cartilage (Brown and Vrahas 1984; Choi et al. 1990; Lotz et al. 1991), the subchondral plate is generally believed to play an important role in juxtaarticular load transmission. This is related to the fact that differences between the trabecular structure, hardness and strength of weight-bearing and nonweight-bearing areas are known to exist. There is normally sufficient deformation of the joint configuration under loading to permit a maximal load-bearing surface to form. The increased congruity of joints in older people (Bullough et al. 1973) implies a diminution in the flexibility of such bone. Jacobs et al. (1976) have shown, by the experimental analysis of stress and strain on epoxy models of the human pelvis, that the subchondral bone of the acetabulum transmits the major part of the weight bearing load in the form of membrane stresses from the hip joint to the acetabular rim and then to the cortical shell of the ilium. It also appears that the cancellous bone of the pelvis is stressed to a greater extent under natural physiological conditions. Extrapolation of the experimentally determined stresses shows that, in an individual of 60 kg, the membrane stresses in the subchondral bone attain values of approximately 98 kPa/cm^2, whereas the normal peak compression within the cancellous material only reaches approximately 13 kPa/cm^2. This phenomenon was also observed by Dalstra (1993), who found the stress levels in the cortical shell to be about 50 times higher than in the underlying trabecular bone.

Hvid (1988) claimed that the dense subchondral bone plate serves to distribute the forces transmitted through the menisci and cartilage to the subchondral trabecular bone.

Experimental studies of subchondral changes induced by repetitive impact loading have shown that the subsequent remodelling results in less compliant trabecular bone, thereby transmitting excessive mechanical stress to the overlying cartilage (Radin et al. 1973; Simon et al. 1972; Burr et al. 1985; Radin et al. 1984). Weight bearing is shared by both the articular cartilage and subchondral bone. Shearing stresses are probably concentrated in the region of the tidemark and osteochondral junction, where cleavage would result in horizontal clefts due to differences in orientation of the collagen fibrils and calcified substance. This may be related to the degenerative changes which are seen in osteoarthritic joints.

It is apparently possible to induce initial degenerative changes analogous to human OA by means of experimentally induced alteration of the load bearing. Serink et al. (1977) showed that increased deformation of subchondral bone results in decreased structural support of the overlying cartilage. The cartilage changes could precede, follow or accompany changes in the subchondral bone. Inoue (1981) suggested that, although degenerative changes are generally encountered in both the articular cartilage and the subchondral bone, the initial changes occur around the tidemark and osteochondral junction. Stresses are concentrated at these sites because by the mechanical structure of the cartilage and subchondral bone. An analysis of Ateshian et al. 1994, too, provided support for the hypothesis that failure due to sudden loading, causing shear-stress failure would take place at the cartilage–bone interface.

Clark and Huber (1989) are of the opinion that, judging from the thickness and lamellar structure of the subchondral bone, the entire layer is probably a site of active remodelling. Inoue (1981) observed that the collagen fibrils of the subchondral plate and trabeculae are clearly impacted in individual lamellae. The fibrils cross each other and the mineral substances are similarly aligned. Deproteinized samples show the effects of such bone dynamics as absorption, resting and deposition.

Furthermore, the subchondral bone plate influences the nutrition of the adjacent cartilage (Duncan et al. 1985). Although it is stated that the nutrition of cartilage is facilitated by compression and relaxation during normal joint use and that the influx and expulsion of synovial fluid from the interstices of the sponge-like hyaline cartilage provide the essential nourishment, some authors have suggested that certain of the deeper layers of the cartilage are metabolically influenced by subchondral capillaries (Ingelmark 1950; Ekholm and Norbäck 1951; Trueta and Harrison 1953; Woods et al. 1970; Sokoloff 1969; Clark and Huber 1989, 1990; Berry et al. 1986; Milz and Putz 1994a).

In conclusion, so far as it is known, the subchondral bone plate fulfils both mechanical and metabolic functions and is an active site of remodelling.

1.5
Possible Pathomechanisms Leading to Osteoarthritic Changes

In normal joints, the interface between the articular cartilage and subchondral bone is subject to shear, compressive and tensile stresses (Radin and Rose 1986). With the loss or degeneration of the articular cartilage, the change in mean trabecular orientation to a direction nearly perpendicular to the articular surface may be a response to the predominant compressive and tensile stresses which act directly on the bone at the joint surface. These structural changes, which accompany OA, may ultimately lead to further joint degeneration as the bone stiffness gradients depart significantly from normal. Although osteoarthritic bone adapts functionally to the altered loading conditions, this adaptation may itself accelerate the progressive failure of the joint.

Normal trabeculae have a very smooth surface, probably illustrating a resting phase. In the presence of OA, the cartilaginous framework becomes frayed as far as the tidemark and the osteochondral junction is obscure. The architecture of the subchondral bone trabeculae is also irregular and this may represent an early stage of arthritic change. Radin et al. (1973), examining the structure of degenerated joints which had been subjected to repetitive loading, showed that the cartilage changes were "preceded" by stiffening of the underlying shock-absorbing subchondral bed and that this stiffening was associated with a significant number of trabecular microfractures and callus formation in various stages of healing. Inoue (1981), however, found no microfractures in the subchondral bone.

Wu et al. (1990) performed varus or valgus proximal osteotomy on rabbits, and found that the cartilaginous deterioration corresponded both spatially and temporally to an increase of bone within the joint. This increased turnover of bone 3–4 weeks after the operation, however, was a transient change also found in surgical control animals. Thirtyfour weeks after the operation, the subchondral plate had increased in width and the porosity below the submeniscal articular surface of the overloaded condyle was significantly reduced. The articular cartilage showed obvious degenerative changes locally. On the other hand, in animals with a valgus deformity, no increase in bone density was found in the area of the lateral condyle not covered by the meniscus and serious cartilaginous changes were also absent from this area. The direct correlation between bone and cartilage changes within a single joint suggested a relationship between the two, which is consistent with the assumption that mechanically-induced cartilage change is a focal process.

These data were confirmed by those of Benske et al. (1988), who found that the cartilage overlying sclerotic bone shows signs of advanced OA, whereas adjacent areas above regions of normal bone density are not arthritic. It, therefore, seems that increased bone density is an essential accompaniment of joint destruction. Whereas these authors found that sclerosis is associated with an increased apposition rate, in the varus/valgus models of Wu et al. (1990), the increased bone density could be attributed to an increase in the total forming trabecular surface and in osteonal bone formation in the subchondral plate. The appositional rates for all groups lay between 0.40 and 0.80 µm/day, which is consistent with the normal rate in the rabbit.

1.5.1
Is Osteoarthritis a "Final Common Pathway" ?

In the osteotomy models of Ogata et al. 1977, Reimann 1973 and Wu et al. 1990, the cartilage changes begin with loss of the superficial and transitional zones. Changes in cell shape and disorganization of the cell columns are characteristic of the osteoarthritic cartilaginous deterioration produced in these models and chondrocyte cloning is common. In another rabbit model of mechanically induced OA which employed repetitive impulse loading (Radin et al. 1984, 1973), the cartilaginous change usually began in the intermediate and deep zones, the cellular changes being only secondary.

The impulse-loading model consistently demonstrated increased subchondral and trabecular bone density in association with both the early and late signs of cartilaginous deterioration (Radin et al. 1984; Yang et al. 1989). Increased subchondral thickness and reduced trabecular porosity are features of the varus and valgus osteotomy models. Layton et al. (1988) demonstrated an increase in the bone-volume fraction and reduced trabecular spacing associated with early osteoarthritic changes in the guinea-pig gluteal myectomy model.

Thus, several etiologically distinct animal models of OA have demonstrated the association of bone and cartilage changes in the progress of OA, suggesting that both increased and decreased bone density is an important component of the final common pathway for mechanically-induced joint failure.

Cartilaginous deterioration can take different forms. Different mechanical conditions can underlie the varying etiology of OA, but bone changes appear to be a constant factor in its pathogenesis. Indeed, there appear to be a number of different causes of mechanically-induced osteoarthritis.

The link between joint trauma and OA has been reported for human joints subjected to repetitive traumatic activity (Davis 1988; Hadler et al. 1978; Hellmann et al. 1983; Hunter et al. 1945; Kern et al. 1988; Parniapour et al. 1990) or following a single traumatic event, such as a fracture involving the ankle joint (Jaskulka et al. 1989; Kellam and Waddell 1979; Moller and Krebs 1982). As a result of the employment of magnetic resonance imaging for examining injured joints, evidence is emerging that subchondral bone changes may occur in the absence of fractures identifiable by conventional radiographical means (Deutsch et al. 1989; Mink and Deutsch 1989; Lee and Yao 1988; Stafford et al. 1986; Vellet et al. 1991; Yao and Lee 1988). There is also some indication that these radiographical changes may lead to subsequent cartilage degeneration (Vellet et al. 1991). Laboratory studies have been directed toward the understanding of these events both in vitro (Radin et al. 1973) and in vivo (Dekel and Weissman 1978; Donohue et al. 1983; Radin et al. 1984). Radin et al. (1984) applied repetitive impulse loading to rabbit knees for up to 6 weeks. Deep splits were observed in the cartilage at 6 weeks, but the earliest changes took place in the subchondral bone, before any noticeable involvement of the cartilage occurred. Bone scans and tetracycline labelling revealed increased bone activity before histological or biochemical changes were evident in the cartilage. Thompson et al. (1991) observed cracks in the subchondral bone and calcified cartilage that did not penetrate the surface of the joint after rapid, highlevel loading of the canine patella. The findings of Vener et al. (1992) suggest that failure in acute transarticular compression loading begins in the zone of calcified cartilage/subchondral bone and subsequently involves the deeper subchondral bone and overlying cartilage. This type of injury may contribute to the development of posttraumatic osteoarthritis after intraarticular fracture at high loads that does not, however, result in gross fracture.

It is now clear that OA cannot be adequately studied either in exclusively biological/biochemical terms or in exclusively mechanical terms. A commonly accepted, but unproven, theory states that the condition is primarily a degenerative disease of the cartilage with secondary bone changes. According to Hulth (1993), there are reasons to believe that primary OA is actually a disorder of the entire joint component and consists of a reactivation of the joint cartilage (remnants of the growth cartilage of the joint head in childhood). In contrast to separate physical, chemical and biochemical approaches, Hulth (1993) presented a comprehensive view. Some growth occurs normally in loaded joints in elderly people. When the joint head expands, the cartilage is mechanically injured by stiffness gradients or impaired nutrition. Attempts to heal the wounds and phagocytose the detritus set loose cytokines, enzymes and additional growth factors within the joint cavity. The substances released into the joint, together with the continuous loading, produce all the typical osteoarthritic changes in the cartilage, subchondral bone and the joint capsule, leading to stasis and increased bonemarrow pressure. Growth of the joint end (epiphysis) in childhood occurs in the basal part of the joint cartilage, which is mineralized. Vessels grow into this zone because of the stimulation provided by growth and angiogenic factors present in the mineralized cartilage (Brown et al. 1988, Canalis et al. 1988).

Bone lamellae are, thus, continuously laid down around the vessels, widening the epiphysis in height and width. This ceases in adulthood, when the mineralized cartilage becomes separated from the superficial elastic cartilage by the tidemark. The biochemical events initiating this inactivity are unknown. The growth function, however, may not have come to a complete standstill. Some studies have demonstrated a latent ability to remodel the subchondral tissue when the joint was exposed to increased load (Farkas et al. 1987; Radin et al. 1975; Walker et al. 1990). In the loaded human femoral head, increased remodelling occurs after the age of 60 (Lane et al. 1977). All these events are probably related to the activation of growth factors, but vary according to the type of joint surface (convex or flat).

With advancing age, the number of tidemarks in the femoral cartilage increases (Lane et al. 1980), implying advancement of the front of the hard tissue. The same is typical of OA. Based on these facts, Hulth formulated his hypothesis that people prone to OA are a subset of individuals with calcified cartilage, which in one or more loaded joints begins to produce factors that stimulate growth. This occurs to a greater extent and earlier in life than the physiological joint growth of old age (Lane et al. 1977). The tendency to develop OA is, amongst other things, dependent on sex and other genetic factors. In most cases, there is an advance of the subchondral hard tissue by only a few millimeters (at the expense of the superficial cartilage) toward the joint surface. The growth is, however, sufficient to cause injury to the articular cartilage by depriving the joint of its resilience at peak loading. In other cases, the opposing cartilage (most commonly in the hip joint) becomes nearly congruous, thus impairing cartilaginous nutrition. Primary OA does not occur in the talocrural joint, which may be because the convex joint surface is not part of a long bone and, therefore, does not grow.

It is possible that there is continuity between physiological growth of the joint head in the elderly (which results in increased joint stiffness) and OA itself. The borderline between the normal and abnormal is not always clear cut and it is related to the degree of activity of the growth factors in the mineralized cartilage and those produced by the loaded and maltreated bone trabeculae.

1.6
Factors Regulating Remodelling of the Subchondral Bone

Almost all studies on the structure and mechanical properties of subchondral bone have shown that there are clear differences between weight-bearing and nonweight-bearing areas within a joint and that the stress distribution is responsible for these differences. In areas of heavy loading, the subchondral bone plate is thicker and denser, exhibits greater strength, stiffness and hardness, is metabolically more active and possesses more blood vessels.

As asked by Amir et al. (1992), what initiates the subchondral remodelling? Wolff's law states that the structure of any bone is a function of the mechanical demands made upon that bone. The load applied to any joint can be modified by many factors, such as weight, the extent and nature of exercise, occupation and integrity of neuromuscular control. Metabolic and systemic influences can also affect the subchondral bone, rendering it less able to withstand normal mechanical demands. Early thickening of the subchondral bone trabeculae has been observed to follow compression (Salter and Field 1960). Sclerosis of bone can result from an increased rate of apposition or decreased resorption, an increase in the area of the bone-forming surface (increased osteoid) or increased endochondral ossification of the calcified layer of the articular cartilage. Regional differences in density in the osteoarthritic femoral head are well known to occur. Amir et al. (1992) found that the mean appositional growth rate and the number of osteoclasts present are the same in the high and low load-bearing areas. The differences in density may be the result of differences in the total duration of new bone formation or resorption at different sites. It is also possible that there are more bone-forming trabeculae in areas which are becoming sclerotic. The possibility that increased endochondral ossification of the calcified zone of cartilage plays a part in the production of a dense subchondral bone must also be considered.

Dekel and Weissmann (1978) investigated the effect of overuse combined with axial peak loading on the knee joints of living rabbits. They recorded early and progressive damage to the articular surface, the presence of an increased amount of prostaglandin E in the synovial fluid, a reduction of cyclic adenosine monophosphate (cAMP) in the subchondral bone and late changes, which were consistent with OA. It has been suggested that cAMP may act as a mediator of the mechanical stimuli in bone and cartilage (Rodan 1981; Davidovitch and Shanfield 1976; Harell et al. 1977) which is consistent with OA. These changes were found only in joints subjected to simultaneous overuse and peak overloading. The results suggest that cartilage damage and chemical changes in the subchondral bone are simultaneous and that both are responsible for the eventual degenerative changes. Frictional overuse alone does not seem to be responsible for the development of OA.

Oettmeier et al. (1992), studying the changes in subchondral bone following strenuous training, found that bone remodelling is clearly elevated, providing a greater volume of subchondral bone in some regions. The trabecular bone diameter remained virtually unchanged, but newly formed bony bridges were found in those undergoing training as runners. In spite of an initially intact articular surface, the subchondral bone becomes thickened as a result of increased bone formation caused by impact loading and exercise. The process decompensates if subchondral sclerosis is advanced and cumulative effects, such as the progression of the mineralization front, impair the mechanical properties of the articular cartilage.

The bone responds differently to repetitive 50/ms and 500/ms impact loading, even though the load and the manner in which it is applied are the same (Farkas et al. 1987). A significantly increased bone mass was found after only 3 weeks subjection to a 50/ms load, whereas no increase was found even after 6 weeks of a 500/ms load. Moreover, no subchondral vascular proliferation was observed in the 500/ms load group. Because the load and its manner of application in these two experiments was the same, except for the rate and duration of peak loading, heavy loading or strain rates may promote vascular proliferation within the subchondral bone. Strain rates for the 50/ms impulse loading (0.03/s) are at the upper end of the physiological range (Rubin and Lanyon 1982), while those of the 500/ms load group (0.002/s) are characteristic of the extreme lower end of the range. Consequently, a strain rate at the upper end of the range could be potentially more damaging to the joint, inducing subchondral sclerosis and subsequent cartilaginous changes.

The finding of a 10% increase in the vessel perimeters and bone mass at the articular surfaces of the talocalcaneal joint and the tibial plateau after 6 weeks of a repetitive impulse loading regime (Farkas et al. 1987) supports the view that a relationship exists between increased subchondral bone vascularity, increased bone remodelling and joint stress. A significantly increased bone mass in the subchondral region was found 3 weeks after the initiation of impulse loading, which confirmed previous studies of early bone change (Radin et al. 1984). This bone formation occurred at the osteochondral junction, brought about by the advancement of newly mineralized bone into calcified cartilage and by the filling in of existing vascular spaces and remodelling of the whole bone. The bone mass continued to increase during the 6 weeks duration of the experiment. The increase in the vessel perimeters represents new vascular formation, with resorption cavities tunnelling through solid bone. These observations indicate that impulse loading promotes early vascular change in the subchondral bone and leads to more active bone remodelling, denser bone and stiffening of the subchondral plate. Apparently, the nature of the load applied is also an important factor, influencing the rate at which these vascular changes occur or even whether they will occur at all.

According to Milz and Putz (1994a), the hydromechanics of the situation must also be taken into account. If one accepts that no continuous material boundary between the articular cartilage and the underlying marrow cavity can apparently be detected, but that there is an accumulation of variously shaped spaces in the zone of mineralization, it is conceivable that the function of these spaces is not only nutritional, but that they also form a communication between the fluid spaces of the cartilage and the marrow cavity. Taking into account the views of Cochran (1988), it seems possible that the movement of fluid can convey changes in pressure in the articular cartilage to the deeper layers of the joint component. The effect of such fluctuations could not only influence the extent and nature of the nutritional supply, but also exercise a formative stimulus on the local osteocytes. Such considerations are particularly supported by the investigations of Kufahl and Saha (1990), who were able to show, by means of a mathematical model, that the osteocytes at a depth of two to five lamellar widths could be reached and nourished through their canals by the movement of fluid. They have also tentatively suggested that the absence of such fluid movement could be the cause of the osteoporotic change which follows immobilization. It is, therefore, no idle speculation to suggest that the development and further functional adaptation of an especially thick zone of subchondral mineralization could be initiated and then guided by such a mechanism.

With this as a basis, one must consider to what extent the movement of fluid under loading can cause the subchondral mineralization zone at least for a short time to "swim" on a kind of pillow. This "pillow", which consists of subchondral and marrow cavity fluid, is at first remarkably stable under loading because of the collecting of cartilage fluid in the deeper regions. It would also be in a position to give way like a spring and, thus, reduce the amplitude of the transient-loading peaks and the steepness of their sides in the time-loading diagram.

There is also an electrophysiological side to these events. New functional aspects which have hitherto received scant attention are associated with the findings of Frank and Grodzinsky (1987a, b). They have shown that, on the one hand, the compression of cartilage by the redistribution of fluid produces potential differences and that, on the other hand, the setting up of potential differences can produce mechanical stress in cartilage. The fluctuating pressure, so produced, is proportional to the voltage, the intensity of the current per surface unit and the frequency.

One can, without being too fanciful, assume that the perforated subchondral mineralization zone functions in a similar fashion to the perforated electrodes used by Frank and Grodzinsky (1987a, b). This is possibly the reason why load-dependent potential differences between the layer of cartilage and the subchondral mineralization zone appear. These potential differences may be induced by electrical charge transfers due to fluid flow in the cartilage, whereby the entire subchondral zone of mineralization is also probably involved through its system of branching canals.

Taking into account the recognized osteoinductive action of electric potentials, one could speculate that the cyclic loading of the articular cartilage generates potentials that provide a stimulus for the growth of bone in directly loaded regions of the subchondral plate. Because of the morphology of the subchondral region, this effect is concentrated on the area of bone lying directly below the cartilage, its action falling off with increasing distance from the joint surface. In addition, therefore, to a formative stimulus which is active throughout the entire skeleton, depending on the distribution of mechanical stress,, there exists a further, purely local stimulus that makes it possible for a fine modulation of the reorganization processes in the subchondral mineralization zone to take place. In this way, it is also possible for the characteristic form of the subchondral mineralization zone to differ morphologically from that of the rest of the skeleton, which is essentially determined by the distribution of mechanical stress.

1.7
The Aims of This Investigation

A review of the literature, together with our own studies, establishes beyond reasonable doubt that the physiological distribution of both the structural parameters (e.g., thickness, mineralization, vascular density) and the mechanical properties (e.g., stiffness or hardness) of the subchondral plate may reflect the long-term stresses within a joint surface imposed by the load acting upon it. This leads to the conclusion that the evaluation of any one of these parameters is able to provide information on how a particular joint is stressed. In a more fundamental sense, the inversion of Wolff's theorem (Wolff 1892) that bone adapts to functional demands by remodelling to reflect the distribution of effective stress (Koch 1917, Hayes and Snyder 1981) permits one to derive biomechanical knowledge from the measurement of one of the above-mentioned parameters.

For a number of reasons (e.g., clinical application, monitoring age-related changes or changes due to different types of treatment), it is essential to have a method which can be applied in the living subject. The ideal method should be easy to use, cause minimal inconvenience to the patient, yield good spatial resolution and, of course, ensure a sufficient degree of accuracy. Several authors (for instance, Isherwood et al. 1976; Reich et al. 1976; Ruegsegger et al. 1976; Genant and Boyd 1977) have advocated the use of X-ray CT for determining the mineral content of bone. It has also been shown that values of the mineral content, obtained by CT, can be used to predict such mechanical properties as ultimate and yield strength, elastic modulus, ultimate energy absorption and yield energy absorption and the penetration strength of the subchondral bone plate (Bentzen et al. 1987).

CT units are available today in most hospitals. CT scanning causes minimal discomfort to the patient and the risk involved in using X-rays on (for example) one arm or leg is negligible, since the dose (typically 10 mGy) is delivered locally.

We have, therefore, attempted to develop a method based on CT, for displaying the distribution of subchondral mineralization, that can easily be used with little risk to the living subject. After the development and justification of this method has been dealt with, its application to basic research and its employment in the clinical field will be described.

Aims in detail
1. Demonstration of the pattern of mineralization in all the large joints in a sample population of healthy people of different ages and its comparison with current models of joint mechanics.
2. Analysis of the changes and deviations from the normal of the subchondral mineralization patterns, observed as a result of comprehensible alterations in the mechanical conditions of a joint.
3. Investigation of the course of the changes in the mineralization patterns, occurring during adaptation to altered mechanical conditions.
4. The development of a procedure based on CT, for demonstrating the pattern of subchondral mineralization in the living.
5. Studies to confirm the validity of the method and its use on clinically relevant joints.
6. Explanation of the physical basis on which the Hounsfield values measured in the subchondral region may be analyzed.
7. Outline of the ways in which a demonstration of the mineralization pattern can also be obtained with minimum errors from joints, which are not easily reached perpendicular to the scan axis.

2 Materials

2.1
Material for the Validation of CT Osteoabsorptiometry (CT OAM)

Specimens of joints from the dissecting room in the Department of Anatomy at Freiburg University were used for the various comparative studies. An itemized list of these specimens is given in Table 1.

All the specimens from the Anatomy Department had been fixed for some time in the usual 2–3% formalin solution, the shortest period of fixation being 6 months. The specimens we used had been kept in the fixative for between 1 and 2 years prior to our research programme (Table 2). External fixation and conservation throughout the whole period took place in a 4% formalin solution.

CT measurements of the density, as a function of the fixation time period and the strength of the formalin solution, were carried out by Schmitt and Hübener (1980) on formalin-fixed tissue samples. They showed that a 6% solution is ideal, since after 8 days and again after 4 weeks, no relevant changes occur in the density. The density values of the samples deviate from those of the initial unfixed specimen by an average of ±3 HU. Certain inhomogeneities in the density, appearing after 24 hours, ought to be mentioned here; they are probably due to the liquid flowing through and can be found even in the central sections, where the density is partially raised. Homogeneous fixation can be achieved by leaving the specimen in the fixative for a longer period.

Fixing the tissue sample in a solution of 3% formalin, as we did, produces a slight reduction in the CT density, even within 24 hours, and inhomogeneity is clearly detectable within the specimen. After 8 days, the samples are, without exception, homogeneously fixed and the CT density is reduced in comparison with measurements taken from the unfixed sample by some 5 to 15 HU. The controls show, after 4 weeks, no significant difference from those examined after 8 days.

Our fixation method, thus, results in a density discrepancy of about 0.5%, since stages in density of up to 200 HU are represented in our investigations. However, because no determination of absolute values, but rather the visualization of differences in concentration is required for a comparative study, this source of error can be ignored both because of it small size and because it is the same for all specimens examined.

Table 1. Age, sex and side of the specimens used

	Specimen	Age (years)	Sex	Side
1. Material used	Shoulder 1	89	M	Right
for comparing X-ray	*Shoulder 2*	75	F	Right
densitometry with CT OAM	Shoulder 3	83	F	Left
	Shoulder 4	92	M	Right
	Shoulder 5	79	F	Left
	Knee 1	72	F	Left
	Knee 2	86	M	Right
	Knee 3	88	F	Right
	Knee 4	60	M	Left
	Knee 5	77	M	Right
	Ankle joint 1	89	M	Right
	Ankle joint 2	80	F	Right
	Ankle joint 3	61	M	Left
2. Material used	Shoulder 1	68	M	Left
for comparing absorption values	Shoulder 2	73	F	Right
with the calcium concentration	Shoulder 3	87	M	Left
	Shoulder 4	81	F	Left
	Shoulder 5	60	M	Right
	Knee 1	78	F	Left
	Knee 2	92	F	Left
	Knee 3	86	F	Right
	Knee 4	60	M	Left
	Knee 5	81	F	Right
3. Material used for comparing primary	Knee	73	M	Right
and secondary sections	Elbow	78	F	Right
4. Material used for comparing tangential	*Knee 1*	*68*	*F*	*Right*
thin sections with axial sections	*Knee 2*	*75*	*M*	*Left*
	Tibial plateau	*72*	*F*	*Right*

CT OAM, computed tomographical osteoabsorptiometry; *M*, male; *F*, female

Table 2. Composition of fixative

10–12 l Distilled water	600 cm^3	Formalin (39.8 %)	=	1.9 %
(depending upon the amount	300 cm^3	Phenol (100 %)	=	2.4 %
of fat present)	50 cm^3	Incidin	=	9.4 %

2.2
Materials Used for CT OAM

The following samples formed the basis for CT datasets, each of which was constructed according to the position of the joint surface and size in sections in different planes and of different thickness.

2.2.1
Spine Samples

5 CT datasets (2-mm slice, transverse plane) of the lumbar spine region from healthy men (30–35 years) and 3 CT datasets (2-mm slices, transverse plane) of the thoracic spine of dissecting-room specimens with a moderate scoliosis were studied.

2.2.2
Shoulder Joint Samples

Eighty-five CT datasets (2 or 4-mm slices, transverse plane) from patients from the Department of Shoulder Surgery, University Hospital, Munich (68 males, aged 18–56 years, 17 females, aged 18–53 years) were used. In 70 patients, the contralateral side was healthy and served as a normal control.

The diagnosis included different forms of instability (ventral, ventro-inferior, ventro-dorsal, dorsal and multidirectional), rotator-cuff rupture and arthrosis. Eleven CT datasets (2-mm slices, transverse plane) from highly trained male gymnasts(aged 16–29 years), who had pursued an intensive training program for at least 5 years were also studied..

2.2.3
Elbow Joint Samples

Thirty-six CT datasets (2-mm slices, sagittal plane) of dissecting-room specimens (17 males, 19 females, aged 58–97 years) were studied.

2.2.4
Radiocarpal Joint Samples

Nine CT datasets of both wrist joints(2-mm slices, sagittal plane) from healthy persons (7 males, 2 females, aged 24–25 years) were used. One wrist joint had to be excluded from the study because of movement artifacts during the CT investigation.

Two CT datasets (2-mm slices, sagittal plane) from patients with a badly reduced fracture of the distal radius (1 male, aged 59 years, 1 female, aged 53 years) and 11 CT datasets from patients suffering from Kienböck's disease (7 male, 4 female, aged 19 to 60 years) were also studied

2.2.5
Hip Joint Samples

Twenty CT datasets (slice thickness 2–4 mm, transverse scans) from subjects who had fractured the pelvis on the contralateral side (aged 18–90 years) and ten (slice thickness 1 mm in the acetabular roof, 2 mm below, transverse scans) from patients with a hip dysplasia (3 males, 7 females, aged 20–48 years) were studied.

2.2.6
Femorotibial Joint Samples

Five CT datasets (1-mm slices, transverse plane) from healthy persons (5 males, aged 24–36 years), two (1-mm slices, transverse plane) from patients with genu valgum (2 males, aged 37 and 45 years) and 14 from patients with genu varum who had undergone a correction osteotomy (4 males, 10 females, 45–72 years) were used. In these patients, the first CT investigation was undertaken preoperatively and the second investigation 1 year after the operation

2.2.7
Animal Studies of the Femorotibial Joint

In each of the 3 following studies, the right knee joint served as control. The CT investigation was carried out 1 year after operation. Six CT datasets (2-mm slices, sagittal plane) from adult merino sheep, in which the anterior cruciate ligament (ACL) had been reconstructed by a patellar tendon graft in the left knee joint were studied. Six CT datasets (2-mm slices, sagittal plane) from adult merino sheep, in which the inner meniscus had been removed in the left knee joint, were analyzed. In addition, six CT datsets (2-mm slices, sagittal plane) from adult merino sheep, in which the inner meniscus was replaced by an autogenic patellar tendon and fascia lata composite graft in the left knee joint, were also studied.

2.2.8
Femoropatellar Joint Samples

Twenty CT datasets (4-mm slices, transverse plane) from patients with a retropatellar pain syndrome (11 males, 9 females, aged 16–65 years) from the University Hospital of Innsbruck were studied.

2.2.9
Ankle Joint Samples

Twenty-three CT datsets (2-mm slices, coronal plane) from healthy persons (14 males, 9 females, aged 23–87 years) were also studied.

3 Methods

3.1
X-Ray Densitometry

Each specimen was sawn into 2-mm transverse or sagittal parallel sections. Contact X-ray films were made from these sections using a mammography apparatus with molybdenum tubes and filtering at 17 keV. The radiographs were standardized using an aluminum–alloy calibration wedge placed on each film, adjacent to the bone sections to be evaluated. These X-rays were then read through a camera system into an IBAS 2000 Zeiss computer, which, with the aid of a digital-image processor, subdivided the continuous gray value distribution of the film into discrete regions, each with its gray value (Schleicher et al. 1980). In order to grade the bone absorption into five steps, the region of optical density between the unattenuated radiation and the radiation attenuated by an aluminum step of 1.2 mm was divided up at equal intervals. By allotting a color to each gray region, a contour map was produced, in which areas of equal density were given the same color.

As a further development of the method introduced by Schleicher et al. (1980), we have been able to create a surface representation of the subchondral mineralization by measuring the density values of each section at a depth of 1.5 mm and transferring them to a surface contour map of the joint. This allows the distribution of the material within the joint surface to be taken in at a single glance.

3.2
CT OAM Used to Demonstrate the Patterns
of Subchondral Mineralization in the Living Subject

The osteometric values are based on conventional datasets which were obtained with various types of CT apparatus.

3.2.1
CT OAM with a Radiotherapy Planning Computer

CT OAM was originally developed on a radiotherapy planning computer. The densitometric evaluation (Fig. 2) of the CT scans, according to the Hounsfield density scale was carried out on the radiotherapy planning computer EVADOS (Siemens, Erlangen, Germany). In addition to the hardware, this system includes the two software programs EVA-1 and SIDOS-TELE 9.4D. As a first step, the individual scans were processed with EVA-1 and the relevant articular surfaces ("regions of interest" - ROI) were enlarged until they filled the entire screen.

Depending on the initial magnification of the CT image, the resolution of the picture lay between 0.1 and 0.8 mm. When the direction was perpendicular to the surface, the resolution was that of the section thickness, i.e., 2 mm.

The next step was conversion into the SIDOS-TELE format. This software first produced a histogram analysis, for which the frequency of the HUs of a ROI was plotted as a function of the density values. Then, a routine designed to seek out the boundaries of density regions was initiated. Isodensits were given for the selected HUs using the SIDOS-TELE subroutine; isodensits being contour lines of equal density, separating regions of greater from those of lesser density. Only regions with more than 20 pixels were recognized, that means, areas of more than 3–12 mm², depending on the scale of the 264 \264 pixel matrix. We chose the following five density regions uniformly for all specimens:

250–499 HU; 500–649 HU; 650–799 HU; 800–999 HU; 1000 HU

From this, one obtained five isodensits. The resulting pictures were then subjected to image analysis. They were first transferred into a computer (IBAS 2000, Zeiss) via a television camera. The various density regions were then represented visually in a false color system, whereby the zones lying between the demarcating isodensits each had its own color.

In order to project the subchondral density onto the joint surface, the density values of each section were – as with the X-ray densitometry described above – measured at a defined depth of 1.5 mm and transferred to a template, whereby a representation of the subchondral density distribution over the entire articular surface could be made.

3.2.2
CT OAM with an X-Ray Computer Tomograph

The next step was carried out in cooperation with the Department of Radiology at the Central Surgical Hospital, Munich. With the help of a program integrated into the CT procedure, it was possible to produce a three-dimensional reconstruction of the density distribution of the subchondral mineralization directly on the tomograph (Fig. 3).

After removing one of the joint components from the CT scans by means of an editing program, a three-dimensional reconstruction of the other component remained, which was seen by the observer in surface view.

Further processing of the scans by means of the editing program left only the subchondral region of the joint surface. The single density regions (200–399 HU, 400–599 HU, 600–799 HU, 800–999 HU and HU) were next assembled separately by the reconstruction procedure described above, so that the total picture then consisted of six images put together.

These images were then transferred, via a camera, into an image analyzer, which provided the single-density images with false colors and then, finally, projected all of them one upon the other.

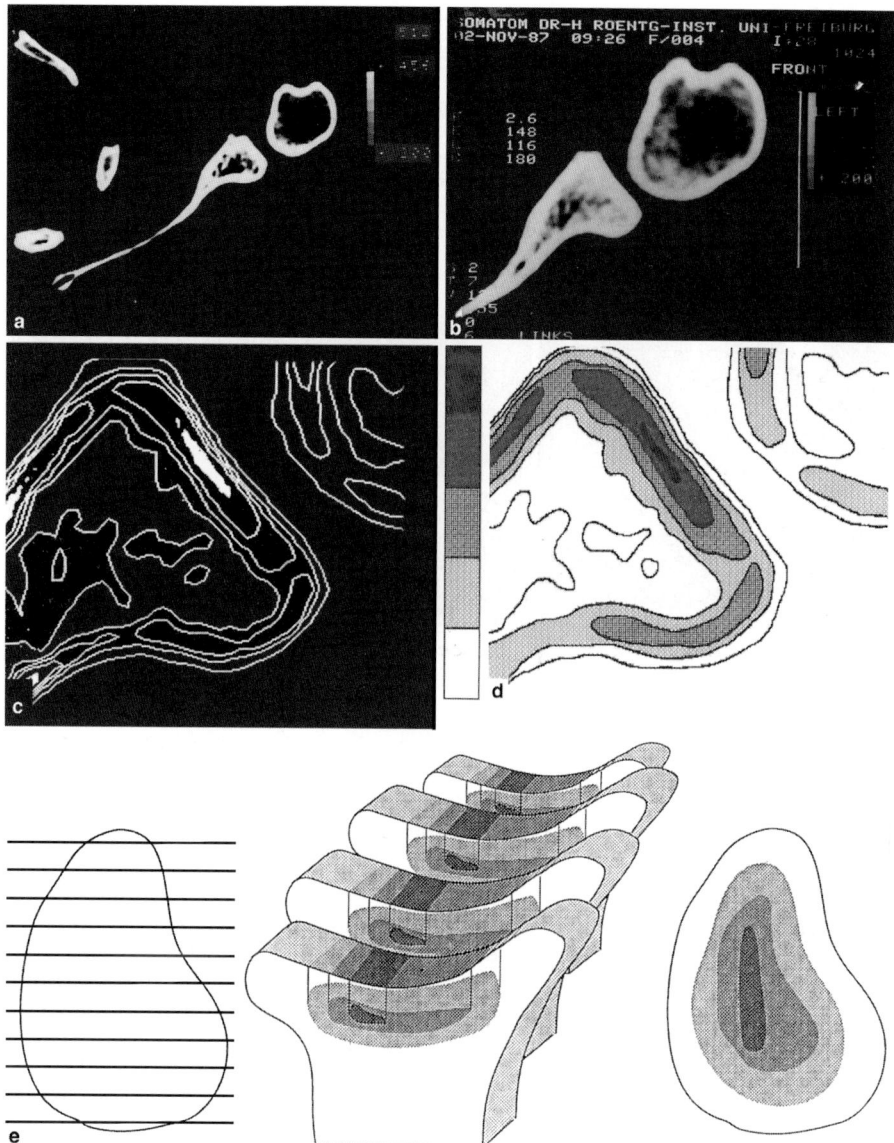

Fig. 2a–e. Method of CT-OAM by means of a radiotherapy planning computer. **a.** Axial CT-scan of a right shoulder joint. **b.** Magnification by means of the program EVA1. **c.** Isodensities in the subchondral region of the glenoid cavity. **d.** False color display (*black* = zones of highest density, *white* = zones of lowest density). **e.** Method of producing density maps from single sections

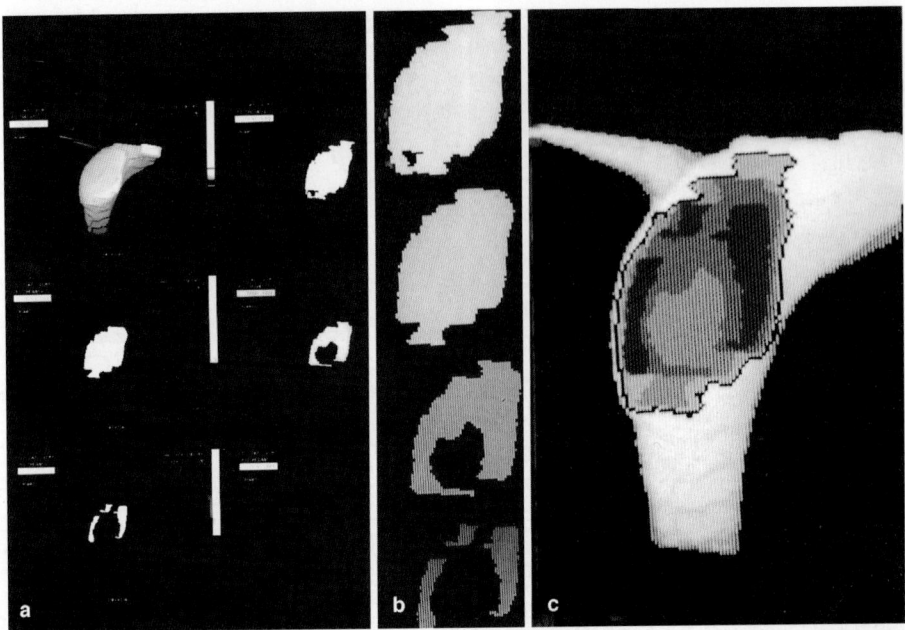

Fig. 3a–c. Method of CT-OAM by means of 3-D reconstruction. **a.** Images of the scapula (lateral aspect) and the different density zones resulting from 3-D reconstruction. **b.** Method of producing density maps in a false color display. **c.** Resulting image

3.2.3
CT OAM Processed by Means of the Software ANALYZE

According to the current procedure, the CT datasets are read into a workstation (IBM RISC System/6000) for further processing (Fig. 4). A three-dimensional picture of the joint surface is then constructed by means of the software program ANALYZE. After additional selective representation of the subchondral bone plate in single sections, the density distribution is processed by "maximum-intensity projection", so that for every image-point, each maximum density value of the underlying bone plate is projected onto the surface. The different Hounsfield stages are classified into steps of 100 or 200 HU and presented in a false-color diagram. Finally, the two pictures are projected one upon the other to arrive at reproducible color-coded mineralization patterns of the joint surface. The zones of greatest density are shown in black or red and less dense zones in green and blue (in black-and-white photographs, the highest density is displayed in black, followed by different gray steps, the lowest density being shown in white).

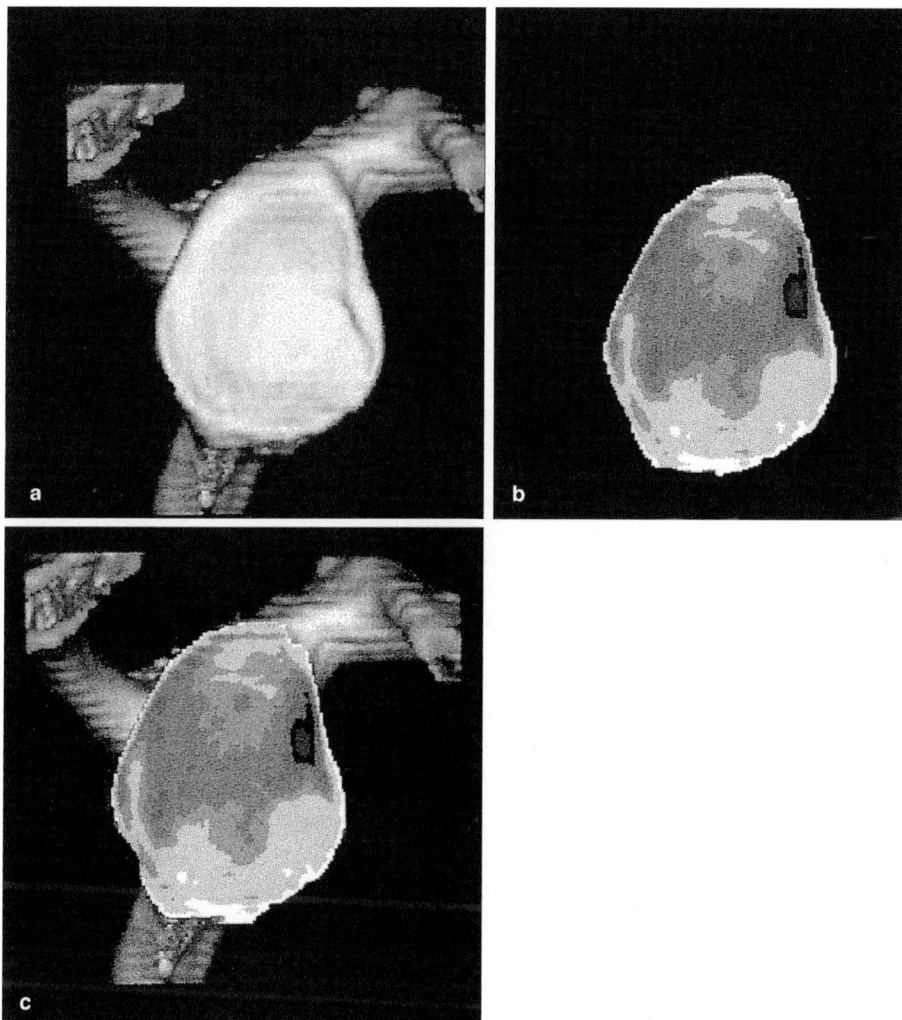

Fig. 4a–c. Method of CT-OAM by means of the Software ANALYZE. **a.** 3-D reconstruction of the scapula (lateral aspect). **b.** Density map of the articular surface of the glenoid cavity by means of 3-D reconstruction and a maximum intensity projection. **c.** Superimposition of both pictures

3.3
The Production of Secondary Sections

If one takes a series of contiguous sections composed of the usual pixels, the columns of the single sections, when placed one upon the other, represent an unbroken scanning of the physical volume in volumetric elements. At the level of the section, the dimension of a voxel in the x and y direction is the same as in the pixel and the dimension in the third direction (z-axis) corresponds to the slice thickness used. By interpolating the orthogonally arranged volumetric elements, it is possible – depending upon the evaluation program – to obtain any section that is required (multiplanar reconstruction). Secondary frontal or sagittal sections may be obtained in this way, although, owing to the partial volume effect, only thin layers produce a satisfactory result.

3.4
Dual-Energy QCT with Basis Material Decomposition

The specimens were examined by means of the dual-energy technique (Fig. 5) with a Somatom DRH scanner (Siemens, Erlangen, Germany). The generation of the material decomposition tables and the data evaluation were carried out on an independent viewing console with an additional array processor. The scanner was equipped with a prototype apparatus for rapid kVp switching. The tube voltage was switched between high and low kVp values from pulse to pulse and, thereby, dual-kVp data were acquired in a single scan. 125 kVp was used for the high-energy spectrum and 85 kVp for the low-energy spectrum. A reference phantom with a small cross-section, comprising two stable plastics, a water-equivalent standard and a bone-equivalent standard, containing 200 mg hydroxyapatite/ml was used.

From each of the dual-energy scans, normal CT images at 125 kV and 85 kV and calcium density images were reconstructed. A matrix size of 512 ⟍512 was chosen.

3.5
Methods of Achieving Standardized Evaluation and Quantification of the Mineralization Patterns

3.5.1
Localization and Displacement of the Maxima

A description of the localization and occasional displacement of the mineralization maxima following a surgical procedure is achieved by establishing a defined subdivision of the joint surface, a separate grid which is optimally "tailor-made" to suit the local conditions being used for each joint.

a. Rectangular grid. The outer limits of the grid are determined by constructing tangents to the most peripheral points of the joint surface. This rectangle is marked by defined subdivision into equidistant sections. In this way, different joints can be compared irrespective of differences in their original size.

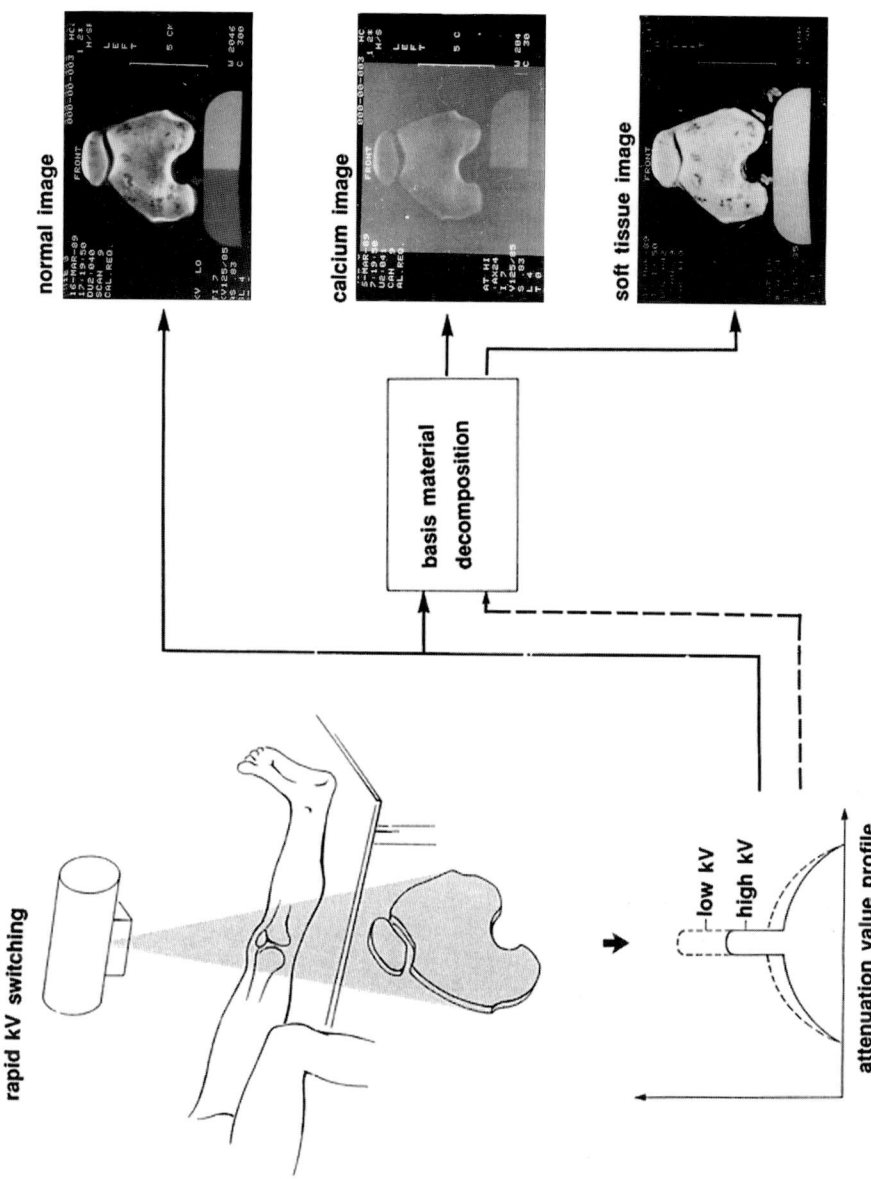

Fig. 5. Principle of dual energy quantitative computed tomography with basis material decomposition

b. Circular grid. A coordinate system, subdivided into three concentric circles is orientated on the geometrical center of the tibial plateau, which lies between the intercondylar tubercles. This divides the surface into quadrants (lateral and medial anterior and posterior compartments), each of which is, itself, subdivided by the circles into areas 1 to 3. The position of the density maxima and, where applicable, their displacement can thus be quantitatively described.
c. Determination of the center of gravity of the surface. The position of the density maxima can also be described by determining the center of gravity of each zone in terms of the defined orientation of the coordinate system followed by a statistical evaluation.

3.5.2
Degree of Mineralization

In order to quantify the total mineralization of a joint surface, the number of volumetric elements for each density step is determined and the percentage of the joint surface which it occupies is then calculated. The mean value of all the included HUs is also determined.

3.5.3
Extent of Changes in Depth

The CT data are transmitted to an IBM workstation in the program ANALYZE via a SUN SPARC 2 station. In editor mode, the subchondral part of the joint surface is depicted in each single section. The proximal limit of the 3D dataset (in the tibial plateau, for instance) begins with the first pixel encountered in the intercondylar protuberance which has a density of over 200 HU and the lower limit at 18 sections below the upper limit. By means of a C program, all the pixels, which are held in binary form in the 3D dataset, are individually evaluated within their slices (18 layers per knee joint) and then expressed in steps of 100 HU from 201 to 1000 HU. The preoperative and postoperative data, so obtained, are finally evaluated by the Wilcoxon matched-pair signed rank test, using the statistics program SPSS.

4 Validation of CT OAM

M. Müller-Gerbl, N. Hodapp

4.1
Comparison with Conventional Procedures

In order to be able to assess to what extent CT OAM actually fulfils its claims and to what errors it is liable, it is first necessary to examine more closely the physical aspects of the radiological methods used in identifying the mineral salts.

4.1.1
The Physical Background to the Assessment
of the Mineral Content of Bone Tissue by Means of the X-Ray Densitometry
of Thin Sections and Using CT or Dual-energy (DE) QCT

The interaction of photons with matter within the quantum energy region of 10–200 keV can result in the following effects:

a. Coherent Scattering (Rayleigh or Thomson scattering). When photons collide with stable outer-shell electrons the whole atom can, under certain circumstances, absorb the energy of those photons. This energy is then totally re-emitted during the short-term collective oscillation of the outer-shell electrons.

b. Incoherent Scattering (Compton scattering). When a photon interacts with outer-shell electrons, it behaves as if it were a particle having no resting mass. The principle of conservation of energy and momentum applies as in particle collision. At collision, part of the energy of the photon is transferred to the electron. After collision, the photon has a longer wavelength (lower frequency) and its angle of dispersion is altered.

c. Photoelectric Effect (Photoeffect). The photoeffect takes place in association with electrons of the inner shell. The energy of the photon has to be greater than the binding energy of the electron. The remaining energy is transferred to the electron as kinetic energy and the photon is destroyed.

The probability of any one of these interactions occurring depends essentially on two parameters: the atomic number of the participating elements and the energy of the photon. That is to say, the extent to which the single interaction shares in the interaction as a whole depends on the nature of the incident radiation and on the composition of the material. The share taken by the single interaction is characterized by the so-called effective cross-section (atomic attenuation coefficient). The effective cross-section has the dimensions of an area and is, therefore, a measure of the "size" of the interacting elements, which is recognized by the photon during each interaction. The atomic attenuation coefficient for the photons of a given energy and for a given interactive element (atom) is made up of the cross-sections of the single interactions. In our case,

$$\mu_a = \sigma_k + \sigma_c + \tau \tag{1}$$

where
μ_a = atomic attenuation coefficient
σ_k = effective cross-section of coherent scattering
σ_c = effective cross-section of Compton scattering and
τ = effective cross-section of the photoeffect

A few examples of the effective cross-section are shown in Table 3. With higher energies, the effective cross-section becomes smaller; another illustration of how the conductive capacity increases. Differences in dependence on the nature of the interacting elements become less as the amount of energy rises. Whereas at 15 keV the contributions of hydrogen and carbon can, as opposed to that of calcium, be ignored (difference in the order of size $\sim$ 100), this is reduced at 60 keV to » 10; that is to say, mammograms give "true" images of the calcium, whereas with CT, the influence of the other interacting elements has to be taken into account.

With macroscopic observations, it is advisable to use the mass absorption coefficient rather than the atomic effective cross-section. The correlation is expressed by:

$$\frac{\mu}{\varrho} = \sigma * \frac{N_A}{M} \tag{2}$$

where
$\dfrac{\mu}{\varrho}$ = mass absorption coefficient
σ = atomic effective cross-section
M = molar mass
N_A = Avogadro's number

Table 3. Examples of the effective cross-section (unit = 10^{24} cm^2/atom) for the energy used for mammography (15 keV) and for CT of moderate energy (60 keV) (Johns and Cunningham 1983)

Energy		Hydrogen	Carbon	Calcium
	σ_k	0.0194	1.958	37.54
15 KeV	σ_c	0.6095	3.023	8.076
	τ	0.0011	9.77	1911
	Σ= 0.63		Σ= 14.75	Σ= 1956
	σ_k	0.0013	0.1957	4.39
60 KeV	σ_c	0.5443	3.108	9.93
	τ	–	0.0941	29.06
	Σ= 0.54		Σ= 3.40	Σ= 43.38

If atoms of different kinds are contained in an absorber, the mass absorption coefficient of this mixture is calculated as the sum of the mass absorption coefficients of its individual components:

$$\left[\frac{\mu}{\varrho}\right] = \left[\frac{\mu}{\varrho}\right]_1 * \frac{m_1}{m} + \left[\frac{\mu}{\varrho}\right]_2 * \frac{m_2}{m} + \dots \tag{3}$$

$\left[\dfrac{\mu}{\varrho}\right]$ = the mass absorption coefficient of each component and

$\left[\dfrac{m_i}{m}\right]$ = the mass quota of each component

4.1.2
The Basis of X-Ray Densitometry

In his contributions to functional anatomy and to the causal histogenesis of the locomotor apparatus, Pauwels (1955) came to the conclusion that the local density of cancellous bone was closely proportional to the distribution of the stress applied in his model. His description was based on an estimated comparison of the structural density of the proximal part of the femoral diaphysis, as seen on X-ray with an assessment of the local stress obtained from an photoelastic model (isochromates). Similar research by the same author (Pauwels, 1963) on the elbow joint revealed the same relationship between local stress shown in the experimental model and the density of the subchondral bone, which led him to coin the expression "incorporated field of stress" (verkörpertes Spannungsfeld). Kummer (1962) came to a similar conclusion. Both assumed from this that the strength of a bone is fairly accurately reflected by the X-ray density.

Knief (1967a, b) quantitatively measured the density distribution of the bone substance at the proximal end of the femur by X-ray densitometry, showing, at the same time, that the stress distribution closely approximates that found in the optical model. Finally, Schmitt (1968), Amtmann and Schmitt (1968) and Amtmann (1971) confirmed the relationship between the radiological density of bone and its strength. Konermann (1970, 1971) developed a photographic method for measuring the X-ray density of bone, by which regions of density are bounded by contour lines ("equidensits"). The density distribution can be read directly from the summation images and can be quantified by means of an aluminum-alloy calibration wedge exposed to it. In 1980, Schleicher and his team further developed the method of equidensits in which the film is scanned and its degree of optical density measured by means of an image-analyzing apparatus.

The great disadvantage of all these methods is that determination of the bone density is only possible from cut specimen sections.

4.1.3
The Basis of CT OAM

Since CT OAM is, itself, based on computer tomography, it is first necessary to briefly consider this basis.

CT, which was developed in 1973 by Hounsfield, is a method by which transverse sections can be depicted radiologically. The construction of the image depends on processing the absorption measurements in a computer as calibrated digital values from which a density scale is obtained.

The image-building principle here can be regarded as an additional development in measuring photon absorption with a radiation detector. The absorption profile of a transverse section through a body, measured in a number of different directions, can be determined in volumetric units of the irradiated layer, by mathematically analyzing the various absorption values of the linear absorption coefficient. The CT picture results from the representation of the attenuation values of the single volumetric units on a surface grid, so that the values obtained can be arranged as gray steps on the monitor. Depending on the density of the radiation bundle, pictorial elements (pixels) and volumetric elements (voxels) are arranged according to their various densities.

The CT scale of measurement lies between -1000 HU for radiation absorbed in air and 0 HU for its absorption in water. The Hounsfield scale is open at the top end. Bone tissue is associated with values of 200 HU and upward.

Commercially available CT equipment works with a boundary energy of radiation between 80 and 140 keV. The filters consist of aluminum sheets, sometimes combined with 0.1 to 0.4-mm copper sheets (McCullough 1982). The mean photon energy lies between 60 and 80 keV.

4.1.4
Conventional X-Ray Densitometry as Applied to Physical Sections Compared with CT OAM

In order to determine whether CT OAM can produce the same results as those obtained with conventional X-ray densitometry, both methods were used to evaluate the distribution of subchondral mineralization in the same specimens and the results were compared. The investigation was carried out on five shoulder joints, five knee joints, three sacroiliac joints and three ankle joints.

To enable a comparison to be made between X-ray densitometry and CT OAM, it is essential that the units in both procedures are the same. The correlation between the X-ray densitogram and the Hounsfield representation is, therefore, achieved as follows. The subdivision of the bone absorption into five zones by X-ray densitometry was carried out by dividing, into equal intervals, the region of optical density between unattenuated radiation and that radiation attenuated by an aluminum absorber of 1.2 mm. The values of the gray scale for the boundaries between the five chosen regions were first recalculated in terms of the thickness of aluminum layers and then equated with the attenuation properties of a standard bone layer, with a constant connective-tissue content and variable hydroxyapatite concentration.

$$(\mu)_{AL}^{L} * X_{AL} = \left(\frac{\mu}{\varrho}\right)_{CT}^{L} * C_{CT} * X_{K} + \left(\frac{\mu}{\varrho}\right)_{H}^{L} * C_{H} * X_{K} \tag{4}$$

where

$(\mu)_{AL}^{L} = 16.393 \; \frac{1}{cm}$ linear mass absorption coefficient of aluminum at 17 keV

X_{AL} = aluminum thickness corresponding to regional margin

$\left(\dfrac{\mu}{\varrho}\right)_{CT}^{L} = 1.341 \dfrac{cm^2}{g}$ mass absorption coefficient for connective tissue (without fat) at 17 keV

$\left(\dfrac{\mu}{\varrho}\right)_{H}^{L} = 11.48 \dfrac{cm^2}{g}$ mass absorption coefficient for hydroxyapatite at 17 keV

C_{CT} = connective-tissue concentration
C_H = hydroxyapatite concentration
X_K = 2-mm thickness of specimen section

This equation was used to assign a hydroxyapatite concentration to each regional boundary. In this way, estimations of the bone composition were obtained and the Hounsfield values for them were calculated by means of the following equation:

$$HU = -1000 * 1 - \frac{\left(\dfrac{\mu}{\varrho}\right)_{CT}^{H}}{\left(\dfrac{\mu}{\varrho}\right)_{W}^{H}} * \frac{C_{CT}}{\varrho_W} - \frac{\left(\dfrac{\mu}{\varrho}\right)_{H}^{H}}{\left(\dfrac{\mu}{\varrho}\right)_{W}^{H}} * \frac{C_H}{\varrho_W} \qquad (5)$$

where

$\left(\dfrac{\mu}{\varrho}\right)_{CT}^{H} = 0.203 \dfrac{cm^2}{g}$ mass absorption coefficient for connective tissue (without fat) at 70 keV

$\left(\dfrac{\mu}{\varrho}\right)_{H}^{H} = 0.344 \dfrac{cm^2}{g}$ mass absorption coefficient for hydroxyapatite at 70 keV

$\left(\dfrac{\mu}{\varrho}\right)_{W}^{H} = 0.194 \dfrac{cm^2}{g}$ mass absorption coefficient for water at 70 ke

$\varrho_W = 1 \dfrac{g}{cm^2}$ density of water

C_{CT}, C_W = concentration of connective tissue and water, respectively.

The values for the materials were taken from tables published by Johns and Cunningham (1983).

The consistently correlated single-section images were also systematically compared to determine what percentage of the image points had the same coordinates in corresponding regions of the two pictures. In order to improve the agreement, an attempt was made to enhance the descriptive model of the bone composition. We took into consideration the following additional points.

The absorption of X-rays by bone is determined by its composition. The enhanced model (four-component model) takes into account, not only the hydroxyapatite and connective tissue, but also the fat and water content. Furthermore, the interdependence of the concentration displacement between the components is more clearly differentiated than in the earlier model (two-component model). Numerous estimations of the composition and density are found in the literature (Frercks 1966; Dulce 1970; Heuck and Schmidt 1960; Piatkowski et al. 1985; Gong et al. 1964). In all these reports, more or less constant values (mean values) are given for both compact and cancellous bone. X-ray absorptiometry and CT osteoabsorptiometry have shown, however, that the absorption, even within similar bone sections, can vary greatly. Since, however, both X-ray densitometry and CT OAM only yield a single value for a specific photon energy and the one homogeneous bone region, these methods can determine only a single parameter for the bone substance.

4.1.4.1
First Model (Two-Component Model)

This model was the basis of the first comparison between the contours given by X-ray absorptiometry and CT OAM. Variations in the absorption or Hounsfield values and, therefore, in the density within a single region of bone were attributed exclusively to differences in the hydroxyapatite concentration, the other concentrations of connective tissue, water and fat being regarded as constant. As a further simplification, these components were assumed to have the same density and degree of absorption. The concentration of the unmineralized components was taken to be 1 g/cm^3. This meant that the hydroxyapatite concentration for Hounsfield values lower than 1000 lay between 0.1 and 0.7 g/cm^3.

4.1.4.2
Second Model (Four-Component Model)

As has already been mentioned, no values are given in the literature for variations of the density within a single bone section. A qualitative report by H.J. Dulce (1970), citing Robinson and Elliott (1957), states that analysis of animal experiments shows the matrix and mineral content to always be proportional, and the water content inversely proportional to the density. A better approximation was then attempted as follows: since the density and mineral content are said to be proportional, the values for the connective tissue, water and fat concentrations given by Frercks (1966) were plotted against the hydroxyapatite concentration. This revealed that the weight–volume concentration of fresh bony tissue from the skull, ribs and from the cancellous bone of the vertebrae and femoral shaft did indeed fulfil the requirements laid down by Dulce (1970) and Robinson and Elliott(1957). This is illustrated by the following regression curve (Fig. 6).

From this curve, the total concentration of the bone substance was calculated as a function of the apatite content. The following equation was derived:

$$C_{Bone} = (0.18 + 0.35^*C_A) + (0.73 - 0.44^*C_A) + (0.19 - 0.13^*C_A) + C_A \tag{6}$$

The first three summands of the row represent the substituted coefficients for the connective tissue, water and fat concentrations and the fourth summand represents that of the apatite. The model, therefore, shows a linear correlation between bone density and apatite concentration. A non-linear fit of the unmineralized bone components would produce no significant improvement in the model. There is a small

46

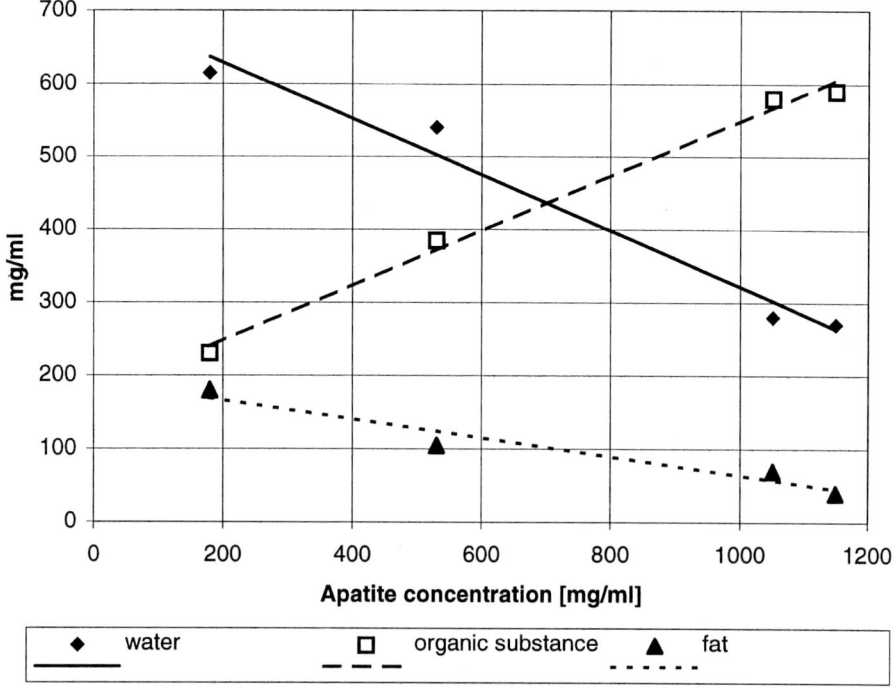

Fig. 6. Regression plot of concentration of the bone substances as a function of the apatite concentration

imperfection in this model, since Frerck's dissertation (1966) could not, for technical reasons, take the cancellous bone into account.

The relationship of the thickness of the aluminum to the hydroxyapatite concentration is analogous to that in model 1 (see Eq. 4). Four components have, however, been included (the mass absorption coefficient of water at 1.25 cm²/g and of fat at 0.82 cm²/g have also been used) and the concentration of each component is expressed as the apatite concentration in the manner described above.

Thus, the following four combinations (analogous to Eq. 4) arise:

$16.4 \text{ cm}^{-1} * n * 0.024 \text{ cm} = (1.34 \text{ cm}^2/\text{g} * (0.18 \text{ g/cm}^3 + 0.36 \text{ C})$ (connective tissue)

$+ 1.25 \text{ cm}^2/\text{g} * (0.73 \text{ g/cm}^3 - 0.44 \text{ C})$ (water)

$+ 0.82 \text{ cm}^2/\text{g} * (0.19 \text{ g/cm}^3 - 0.13 \text{ C})$ (fat)

$+ 11.48 \text{ C}) * 0.2 \text{ cm}$ (apatite)

The scaling of CT OAM values was successfully achieved by converting them to HUs. The relationship between HUs and the mass absorption coefficients is given by:

$$HU = -1000 * \left(1 - \frac{\left(\dfrac{\mu}{\varrho}\right)_1}{\left(\dfrac{\mu}{\varrho}\right)_W} * \frac{C_1}{\varrho_W} - \frac{\left(\dfrac{\mu}{\varrho}\right)_2}{\left(\dfrac{\mu}{\varrho}\right)_W} * \frac{C_2}{\varrho_W} - ... \right) \qquad (7)$$

where

$\left(\dfrac{\mu}{\varrho}\right)_1$ = mass absorption coefficients of the individual components

C_1 = concentration of the individual components and

ϱ_W = the density of water

The concentration values of the participating components had, in our case, a mean photon energy of 70 keV (Table 4)

All values for the substances were again either taken from Johns and Cunningham (1983) or calculated from their tables.

When comparable single sections – CT scan and X-ray images at the same level in the specimen – are placed together, the far-reaching agreement in density distribution is obvious at a glance. Comparison of the density maps for the same joint (Fig. 7) shows, even more clearly, that the results obtained from one joint by each method are virtually identical.

From the systematic comparison of single corresponding sections, we learned that an average of 81.8% of the imaging points of the same density (with five selected degrees of density) always showed the same coordinates in the objective image. 8.4% of the corresponding image coordinates differed by one degree of gray and 9.9% by two or more degrees of gray.

To compare X-ray densitometry with CT OAM, Hounsfield values were classified according to the degree-of-gray values via the calculated hydroxyapatite concentrations. Table 5 demonstrates the results achieved after converting the degree-of-gray values of X-ray densitometry to apatite concentrations with the two-component model.

Table 4. Mass absorption coefficients of the components of bone at 70 keV

$\left(\dfrac{\mu}{\varrho}\right)_{CT}$ = 0.203 cm2/g for connective tissue

$\left(\dfrac{\mu}{\varrho}\right)_{H}$ = 0.344 cm2/g for hydroxyapatite

$\left(\dfrac{\mu}{\varrho}\right)_{W}$ = 0.194 cm2/g for water

$\left(\dfrac{\mu}{\varrho}\right)_{F}$ = 0.188 cm2/g for fat and

Table 5. Conversion of the gray scale (X-ray densitometry) into apatite concentrations (two-component model)

Gray value	Aluminum (mm)	Apatite concentration (g/cm3)
0	0.0	–
51	0.24	0.06
102	0.48	0.23
153	0.72	0.40
204	0.96	0.57
255	1.2	0.75

Table 6. Conversion of the gray scale (X-ray densitometry) into apatite concentrations (four-component model)

Gray value	Aluminum (mm)	Apatite concentration (g/cm3)
0	0.0	–
51	0.24	0.06
102	0.48	0.23
153	0.72	0.40
204	0.96	0.57
255	1.2	0.74

With the four-component model, various thicknesses of the aluminum sheet and apatite concentration values were assigned to the degree-of-gray values (Table 6).

Comparing the Tables 5 and 6 shows that use of the four-component model does not differ significantly from the simpler method. With CT OAM, the two-component model yields the following formula for deriving the Hounsfield values from the hydroxyapatite concentration:

$$HU = 50 + 1770 * C_A \tag{8}$$

where

C_A = the hydroxyapatite concentration

Bearing in mind the characteristic values of the substances as given above and the interdependence between the concentration levels of single components, the four-component model yields

$$HU = 100 + 1560 * C_A \tag{9}$$

49

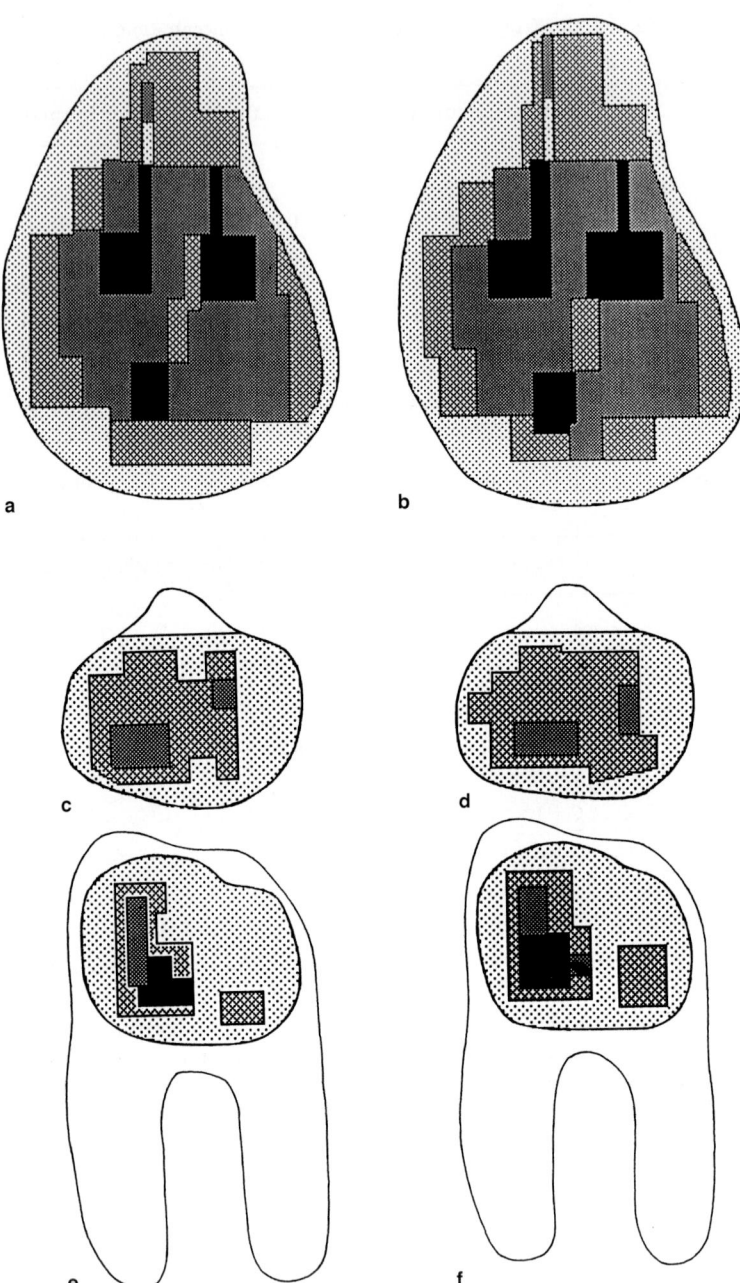

Fig. 7a-f. Density maps of the subchondral mineralization (**a**) in the glenoid cavity (lateral aspect) by means of X-ray densitometry, (**b**) in the same glenoid cavity (lateral aspect) by means of CT OAM, (**c**) in the patella (dorsal aspect) by means of X-ray densitometry, (**d**) in the same patella (dorsal) aspect) by means of CT OAM, (**e**) in the femoral condyles (ventral aspect) by means of X-ray densitometry and (**f**) in the same femoral condyles (ventral aspect) by means of CT OAM

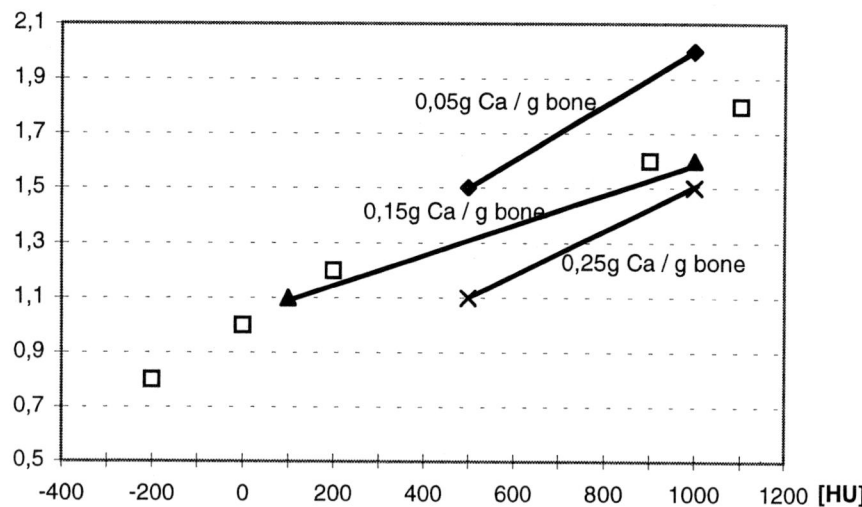

four-component model

[Ca g/ccm]

Fig. 8. Empirical relationship between Hounsfield units (*HU*) and physical density

Table 7. Comparison of the composition of bone resulting from the two models

Gray value	Apatite concentration(g/cm3)	HU model 1	HU model 2
51	0.06	156	194
102	0.23	457	459
153	0.40	758	724
204	0.57	1058	989
255	0.74	1360	1254

If one represents the density-dependence of the Hounsfield number as an ordinary correlation curve of both sizes (density via the Hounsfield values, Schlegel, personal communication), one obtains the curve shown in Fig. 8.

The comparison of the composition of bone (Table 7) shows that differences in the calculated regional boundaries due to the model used can amount to 100 HU.

The reproducibility of the method was determined by sectioning the surface of the same joint on six different occasions after selecting different points in the running time of the CT apparatus and obtaining the results by CT OAM. Comparison of the reconstructed distribution patterns was carried out by image analysis (Vidas/Kontron); the percentage of identical image points and of those varying from one exposure to another were evaluated. Comparison of the surface reconstruction of the subchondral mineralization from six different sets of exposures by image analysis demonstrated an average exact reproducibility of 80% of the points on the joint surface. Eleven percent of the pixels showed a deviation of one density step (out of eight) and a further 9% with a deviation of more than one step. These deviations (about 20%) can be accounted for by a slight tilting or rotation of the surface, so that the orientation of the specimen for each of the tests was not always the same. However, the relative distribution of the subchondral mineralization, which is what one is interested in, is so little influenced that the interpretation of the density pattern is not essentially influenced.

Our findings have shown that X-ray densitometry and CT OAM both lead to the same apatite distribution pattern. The essential difference, however, lies in the fact that X-ray densitometry can only be applied to anatomical sections, whereas CT OAM is also available for living subjects.

The reason why only an 81.7% agreement between the imaging points of similar bone density is reached when using both methods to evaluate the results of comparing similar sections, is, in our opinion, partly because similar sections generally evince a slight positional difference. In other words, the section level in the anatomical specimen and the scan level are not exactly the same. There is also the problem of the variability between X-ray densitometry and CT OAM in the choice of regional boundaries. Further deviations occur because this boundary selection does not exactly correspond between the two techniques, as could be demonstrated in both the two-component and four-component models.

In the models we used, a comparison of the computed regional boundaries shows that variations due to the influence of the models themselves can amount to as much as 100 HU. These models, however, are based on data on bone composition given in the literature; and, because of the wide range of variation found, we could only work with average values. A further difficulty arises in that no specific data on the biochemical composition of subchondral bone are as yet available.

However, much more extensive errors can, during the conversion of regional boundaries, arise from surface variations in the section thickness of the X-ray densitometry specimens.

The fixed boundaries are selected in the light of the overall background. The unequal distances between the Hounsfield intervals certainly cannot be attributed to deviations from the accepted linear relationship between the physical density and the Hounsfield values themselves.

The conversion of the chosen regional boundaries in the two methods (X-ray densitometry and CT OAM) by means of the two-component model also shows that the contribution of the unmineralized fraction is, for X-ray densitometry, insignificant. Nevertheless, our results make it clear that, in spite of the unsatisfactory condition of the mass absorption coefficients (unsatisfactory, because of the difference between the mineralized and unmineralized components when it is the mineralized component that one wishes to represent), the CT OAM values deliver the same results for constructing articular surface maps as does conventional X-ray densitometry. For this reason, it is possible to compare results already obtained that involve several joints (Pauwels 1963; Tillmann 1969, 1971, 1978a, b; Oberländer 1973; Möllers et al. 1986; Putz et al. 1987; Müller-Gerbl et al. 1989, 1990) directly with the findings of CT OAM.

At this point, it is necessary to emphasize that the purpose of CT OAM is to record differences in the density distribution within the subchondral layer of bone. It is not the aim of this procedure to provide – as in the case of the familiar methods of CT densitometry for the diagnosis of osteoporosis, for instance – absolute values in g/cm^3 over a large area. Furthermore, because of the partial-volume effect (examined below in greater detail), CT OAM can only be used with the subchondral or compact layers and not with cancellous bone. It was to make this fact quite clear from the beginning and to avoid misunderstanding, that the name "CT OAM" was selected, since it is concerned with the representation of the absorption distribution within the subchondral bone of the joint surface.

For every joint examined, there are, however, conditions to be observed which relate to the idiosyncrasies of the CT method itself. It is obvious that reliable measurements of the density can only be expected in regions of the program that are free from artifacts, such as those arising from movements of the patient during exposure, for example, or from the proximity of a metal implant.

Further sources of error include:
a. *The partial-volume effect.* Because of the defined layer density of the tomogram, only that part of a structure lying within the layer examined is recorded. During the reconstruction of the image, every voxel is treated as a homogeneous entity. When the material within a voxel varies in density, only a mean value is recorded and this can lead to an artificial representation of unreal dense structures that are not present in the body. This phenomenon may exert an undue influence on the imaging and densitometry of small organs (the adrenal, for instance) or on the peripheral components of organs where the surrounding connective tissue inevitably has a lower density (Hübener 1981). In certain circumstances, the reduction of the partial-volume effect can be achieved by using thinner layers for the scan or by excluding the peripheral density of the organ from the evaluation. To a limited extent, magnification can also help to reduce this error. In any case, the effect is also found when preparing radiographs of anatomical sections, wherever the limiting surfaces of structures of different density are not perpendicular to the plane of the film. This applies particularly to the surface of a joint. Density data from the individual zones in a layer can, therefore, only be obtained if care is taken to ensure that the CT plane is, throughout the whole extent of the section, as far as possible, perpendicular to the joint surface. By avoiding tangential sections, which pass through regions with a steep density gradient, this effect can be significantly reduced.
  By using "maximum-intensity projection" with the latest methods for representing the density pattern, which project the highest density value within the subchondral lamella onto the surface, it can become so small as to be negligible.
b. *Fluctuations due to the equipment.* Internal calibration of the scanner against air is necessary before taking each measurement. Control measurements in a phantom have revealed an unavoidable deviation of 4 HU which, for our purposes, can be ignored.
c. *Effects due to beam hardening.* The main effect of hard radiation can be compensated for by calibration against water. Deviations in the deeper regions of the subchondral bone, where our measurements are made, amount to 4% or less (Kouris et al. 1982). Compared with density intervals of 200 HU, these are again negligible.
d. *Errors due to overlapping with substances of a different atomic number; in this case mineral salts, connective tissue and fat.* Unlike CT densitometry, which deals with

the determination of the mineral salt content of a large section of bone (when diagnosing osteoporosis, for example), the necessity for repositioning the sections as accurately as possible before repeating the examination can be ignored, since the final results of CT OAM are expressed as density maps.

The development of a 3-D process for use with CT OAM offers several advantages. The circuitous route through the radiation-therapy planning computer, by which the isodensits were calculated in the original method, is no longer necessary. With the 3-D reconstruction program, the calculation of the density intervals can be carried out directly by the CT apparatus. This is, in itself, a great advance, because, in addition to our original procedure, the shape of each articular surface can obtained much more quickly.

The application of CT OAM to living subjects naturally raises the important question of radiation exposure. With the latest scanners, however, it lies within tolerable limits. According to Cann (1988), "the radiation exposure of a single CT-examination [is] comparable to less than a 1-month exposure to natural radiation or less than the exposure for a single chest radiograph". The radiation exposure with CT must, as opposed to that with an overall radiograph, be taken into consideration for each single layer. The number of layers, their thickness, the tube voltage and the filtering all contribute essentially to the dosage level.

Taking into account the possible sources of error, CT OAM offers a procedure which has the great advantage of being suitable for use on living subjects. We have, therefore, followed this up by considering several clinical situations in which it can be applied.

4.2
Dependence of the Absorption Value
on the Calcium Concentration

The relationship between the X-ray densitometry (or Hounsfield) values and the composition of the bone was established in terms of the two models described above. A direct method would involve the determination of the calcium concentration and, thus, the apatite concentration by means of DEQCT.

With this in mind, we examined the correlation between those Hounsfield values and calcium concentration values which can be obtained from the same section by DEQCT, together with a breakdown analysis of the basic components.

4.2.1
Fundamentals of DE QCT
with Basis Material Decomposition

Determination of the proportionate content of mineral salts, water-equivalent bone matrix and fat can be carried out with DEQCT (Genant and Boyd 1977; Adams et al. 1982; Kalender et al. 1986). In this way, a raw dual-energy dataset was developed from the breakdown of the CT values from substance-detecting CT images.

In its simplest form, the dual spectrum procedure can make use of weight subtraction; by film subtraction, for instance, or by mixed subtraction with dual photon absorptiometry (DPA = dual-energy photon absorptiometry, Shimmins et al. 1968; Strüter and Rassow 1969). With this method, only substances with high atomic numbers can be imaged. For the determination of other substances, as for the precise quantification of single substances, the breakdown analysis of the basic substance is, in principle, an essential requirement.

We needed two substances which could be sufficiently distinguished by their atomic numbers and, therefore, by the proportions in which their radiation attenuates via the photo- and Compton effects. The two substances selected, i.e., water and calcium, were designated as basic substances.

On the relevant assumption that the attenuation is due only to the photo- and Compton effects, it follows that the attenuation coefficient of any chosen substance can be approximately expressed as a linear combination of these or as the mass attenuation coefficient of the two basic components (Alvarez and Macovski 1976). This is given by:

$$\mu(E) = \varrho_1 \left(\frac{\mu}{\varrho}\right)_1 (E) + \varrho_2 \left(\frac{\mu}{\varrho}\right)_2 (E) \tag{10}$$

As will later be described more precisely, ϱ_1 and ϱ_2 signify the density equivalent of the basic components. A limitation only appears for substances with very high atomic numbers, the marginal energy of which falls within the chosen region of the X-ray spectrum. Thus, the attenuation (the integral μds) that occurs via the absorption path can only be attributed to two basic components for each object of selected composition:

$$\int \mu(E) ds = \left(\frac{\mu}{\varrho}\right)_1 (E)\sigma_1 + \left(\frac{\mu}{\varrho}\right)_2 (E)\sigma_2 \tag{11}$$

Here, σ_1 and σ_2 signify the so-called basic component equivalent densities, which can be expressed as integrals over the local density distribution, where $\varrho_i (r)$ is the density distribution of the region.

$$\sigma_i = \int \varrho_i (r) ds \tag{12}$$

where i = 1.2.

In place of the attenuation value $\int \varrho_i(r)ds$, the integrals σ_1 and σ_2 must be determined in order to be able to calculate attenuation-density images for the two basic components instead of the usual X-ray images.

Evaluation of the mass distribution (mass/area) σ_1 and σ_2 follows from measurement with two spectra, whereby the degrees of attenuation, I_h and I_l, of the spectra for high and low voltage are obtainable from the primary intensities I_{oh} and I_{ol} for each burst of radiation. This yields the following system of equations:

$$I_h(\sigma_1,\sigma_2) = \int I_{oh}(E)^{-\left(\frac{\mu}{\varrho}\right)_1 (E)\sigma_1 - \left(\frac{\mu}{\varrho}\right)_2 (E)\sigma_2} dE \tag{13a}$$

$$I_l(\sigma_1,\sigma_2) = \int I_{on}(E)^{-\left(\frac{\mu}{\varrho}\right)_1 (E)\sigma_1 - \left(\frac{\mu}{\varrho}\right)_2 (E)\sigma_2} dE \tag{13b}$$

in which only σ1 and σ2 are unknown. Using a series of suitable standard measurements, it is possible to produce tables from which, for each pair of attenuation values for high and low energy, a pair of values for σ1 and σ2 can be obtained.

4.2.1.1
Selective Measurement and Imaging of Substances

By analogy with Eq. 10, the two basic components can be regarded as vectors defining a plane in which all other substances can be represented. If only the two basic components appear in the object – soft tissues and calcium, for instance – these can be ideally separated by each basic vector.

This selectivity is, however, in principle, limited if a third substance, the atomic number of which is different from those of the other two, is present. Any third (or more) substance makes a contribution to both basic component vectors or to the associated basic component density image.

A substance with an atomic number (N) which lies between that of water ($N = 7.4$) and calcium ($N = 20$) contributes positively to both images. An example would be phosphorus ($N = 15$).

A substance with an atomic number which is higher than that of either basic component would make a positive contribution to that with the lower number. Iron ($N = 26$) would, therefore, appear with a high positive contrast in the calcium image, but a negative contrast in that of water.

A substance with an atomic number lower than that of either basic component would, on the other hand, make a positive contribution to the water image, but a negative one to that of calcium.

In practice, this loss in selectivity is, in most cases, of little significance. Particularly when the atomic number of a substance differs only slightly from one of the basic components, the contribution to the density image of the other substance is small. In measuring the mineral content of bone, for instance, the influence of fat on the density of the calcium image can virtually be ignored (Vetter et al. 1986), since the elements of similar breakdown analyses have a fixed correlation with one another – in apatite or in connective tissue, for example. Furthermore, the size of the contribution made by a single substance to the density image of the basic components can be calculated and the error deducted (Lehmann et al. 1981).

For any one section of the body, the profile of the attenuation values obtained with higher or lower tube voltages differs only slightly in regions with connective tissue. In bone, on the other hand (in a vertebral body, for instance), there is a considerable difference because of the increased attenuation with the low-energy spectrum. The composition of the object measured can, however, be determined from the difference in the attenuation values (Alvarez and Macovski 1976; Kalender et al. 1986).

Images of the substance density can be calculated from the attenuation values measured. Instead of a conventional CT image, representations of the calcium and soft-tissue density can be obtained.

By processing a raw dual-energy dataset, it is possible to obtain conventional CT images with high (kV-high) and low (kV-low) voltages. In conjunction with breakdown analyses of images of calcium (*mat*-high) and connective tissue (*mat*-low), further parameters, such as the electron density images of the effective atomic number and monoenergetic images can be calculated.

4.2.1.2
Calibration of Values Against a Standard Phantom Bone

With QCT, the linear and mass absorption coefficients and the CT density values of bone, expressed as HU, are dependent on the spectral-energy distribution of the continuous X-ray spectrum. This can, because of variations in the type of CT equipment used, lead to marked differences in the measurement of the mineral salts. However, the energy distribution is known, so that, although different intermediate distributions appear, the end results are the same. Because of variations in the attenuation characteristics of the material to be examined – as well as in its shape, the tube voltage, the filters and the hardness of the radiation – a calibration of the CT scale becomes essential. Simultaneous scanning of the object under examination, together with a substance of known physical and chemical qualities to establish calibration of the CT scale, will permit standardization of the attenuation values, regardless of which CT apparatus is used.

A standard bone reference for measuring the mineral salt content by CT should, because of the dependence of the linear absorption coefficient (μ_K) of the bone tissue on the effective atomic number (N_{eff}) and on the bone density (ϱK), conform as much as possible to bone tissue with regard to the effective atomic number and the density as a function of the mineral concentration (C_M).

From this point of view, a good reference standard is provided by an aqueous solution of K_2HPO_4, as first used by Meema et al. (1964). With increasing age of the phantom-bone reference, however, instability of the K_2HPO_4 solution leads to the formation of air-bubbles, precipitation and impurities (Reiser et al. 1985). In order to eliminate these imponderables, a solid phantom (Siemens, Erlangen, Germany) was used. This was developed from the research carried out by White (1978) and was based on polyethylene.

The addition of a small homogeneous mixture of other substances to the basic material gives it aqueous properties and a degree of absorption that differs from water by less than 1% in the region between 30 keV and 3 meV. For the energy band (40–100 keV) used in CT a maximum deviation of <0.2% was measured for the attenuation properties. This reference substance is also the equivalent of water, as regards the Compton effect and photoeffect (Reiser et al. 1985).

The equivalent of bone material is achieved by introducing 200 mg/ml of hydroxyapatite, homogeneously, into one half of the synthetic phantom. This phantom is $400 \times 90 \times 25$ mm in size and is placed beneath the patient or specimen at the same depth as the examination mat.

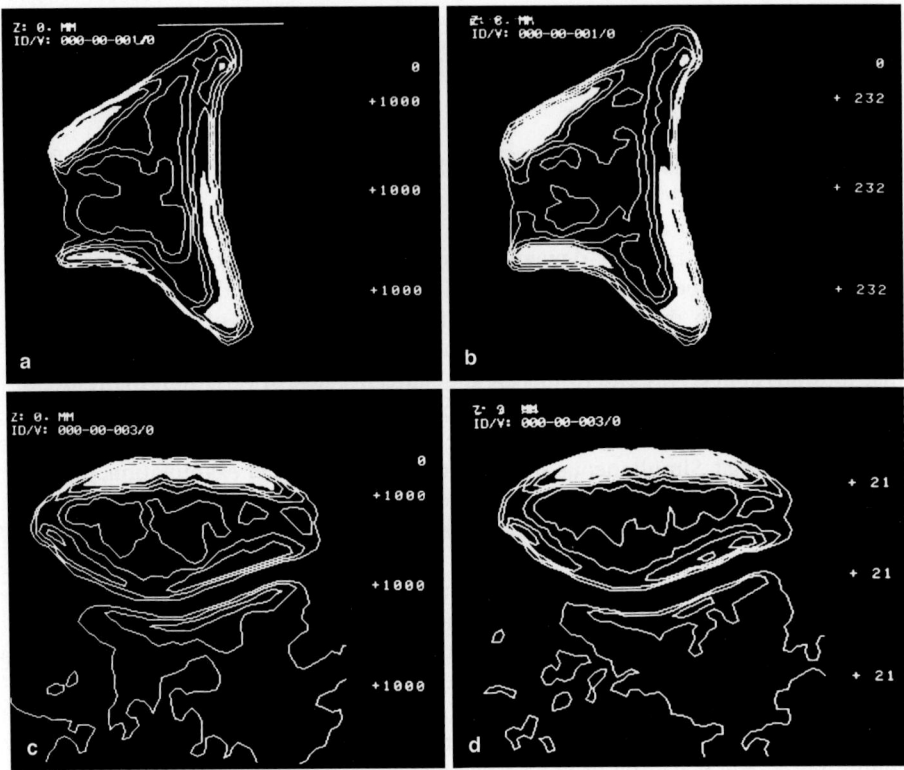

Fig. 9a-d. Isodensity images of the Hounsfield Units (HU) and calcium values in identical sections. **a.** HU image of a right shoulder joint. **b.** Ca image of a right shoulder joint. **c.** HU image of a right femoropatellar joint. **d.** Ca image of a right femoropatellar joint.

4.2.2
Comparison of the Hounsfield and Calcium Values

In a single section from each specimen, both the Hounsfield values at each of 50 selected points in the normal CT images and also the calcium values at identical coordinates in the *mat*-high images were determined. These values were then plotted against each other as Cartesian coordinates and a regression analysis carried out.

In order to obtain both an OAM evaluation of the calcium images and an image projection in the form of isodensits, the regional boundaries corresponding to the HUs were read off from the regression line (Fig. 9).

By plotting those Hounsfield and calcium values taken from the same coordinates in a single section against each other on a Cartesian coordinate system, a linear relationship was obtained between the two parameters (Fig. 10). The correlation coefficients elicited for all ten specimens examined lay between 0.85 and 0.98 (Table 8). A DEQCT- determined correlation of the Hounsfield and calcium values showed no systematic deviation from the linear relationship.

Ca-value

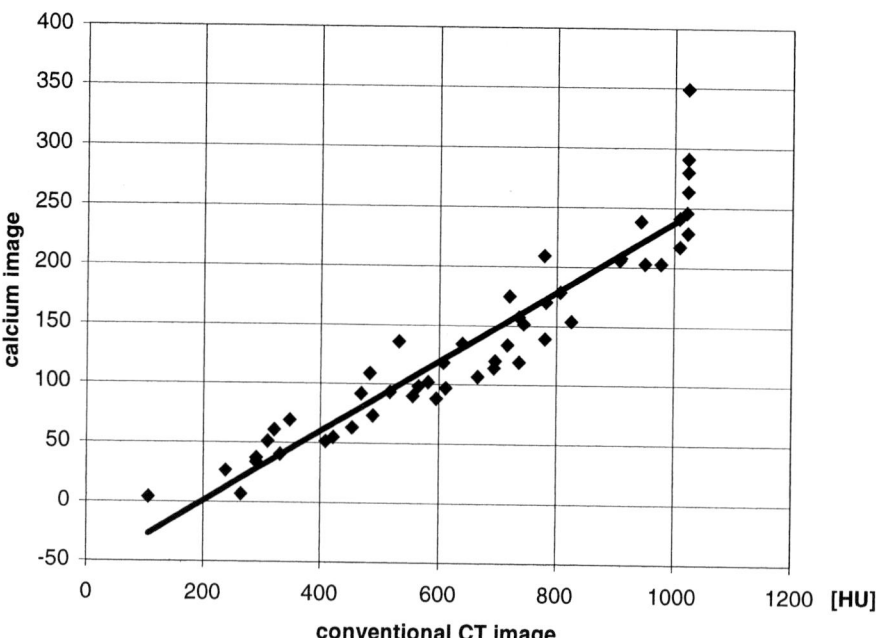

Fig. 10. Regression line for one shoulder joint between HU values (*x-axis*) and Ca values (*y-axis*)

Table 8. Correlation coefficients for the 10 specimens examined

Shoulder joint 1	0.96	Shoulder joint 2	0.90
Shoulder joint 3	0.94	Shoulder joint 4	0.88
Shoulder joint 5	0.85	–	–
Knee joint 1	0.98	Knee joint 2	0.98
Knee joint 3	0.97	Knee joint 4	0.97
Knee joint 5	0.96	–	–

The conclusion, already theoretically reached from our observations of the relevant effects of X-ray absorption, that the first and foremost achievement of CT OAM is the representation of the mineral salt distribution, has been impressively confirmed by DEQCT. The values obtained with this latter method show that, in the subchondral bone region, the normal Hounsfield distribution is not obscured in any way. Since there is no deviation from the linear correlation between the calcium values obtained with DEQCT and the Hounsfield values, it may be assumed that the concentration of the unmineralized component of the bone is systematically dependent on the mineral salt concentration.

Thus, Pauwel's assertion (1965) that the quantitative distribution of the subchondral bone density reflects the principal long-term stress acting on a joint, must be restricted by applying it only to the subchondral mineral concentration.

Taking into account the above-mentioned deviations, which, for our measurements, may be regarded as of little significance, the degree of joint mineralization in anatomical specimens or patients can safely be compared without reference to a phantom. However, it must again be emphatically stated that our methods are not concerned with the calculation of absolute values. They deal, rather, with the representation of differences in the relative concentration within the joint surfaces. To obtain quantified density measurements which remain comparable even after the passage of time, either DEQCT must be used (Adams et al. 1982; Genant and Boyd 1977; Kalender et al. 1986), or the measurement must be made against a phantom reference (Genant and Boyd 1977; Kalender et al. 1986; Meema et al. 1964).

4.3
The Use of CT OAM in Connection with Sections Cut at Other Angles

The position in the body of certain clinically important joints – knee and elbow, for instance – is sometimes restrictive when preparing tangential sections. Since, however, these are essential in order to obtain images which are as orthogonal as possible to the joint surface, we had to find out whether secondarily reconstructed sections could offer adequate relevant data in spite of the partial-volume effect.

CT datasets with a section thickness of 2 mm in the transverse, frontal and sagittal planes were prepared from one knee and one elbow joint. Reconstruction into secondary sections in the frontal and sagittal planes was made from the dataset of transverse sections. All the primary and secondary sections were also absorptiometrically evaluated by means of the CT OAM method described above.

As a first systematic comparison, the Hounsfield values of the appropriate image coordinates of the corresponding primary and secondary coordinates (a total of 11 pairs, 50 imaging points to each pair) were processed in an Evaluskop (Siemens) and transferred to a system of Cartesian coordinates.

As a second comparison, the extent of the area of the individual degrees of density were determined for corresponding primary and secondary sections and the extent of area coincidence was measured.

When comparable single sections are contrasted with one another (corresponding primary and secondary sections), the good agreement between the figures for the primary and secondary sections can be seen at a glance. In the sections compared, the course (expanse) of the density intervals shows a relatively good amount of agreement.

When the density maps for one joint surface (Fig. 11) are compared, they likewise demonstrate that the surface distribution patterns on the map pairings – i.e., one from the primary section, the other from the secondary section – are closely in agreement with each other.

Systematic comparison of the Hounsfield values of the matching image coordinates from the corresponding primary and secondary sections reveals a linear relationship (Fig. 12), with correlation coefficients between 0.983 and 0.996 (Table 9). The additionally determined deviation of the straight slope from its ideal value of 1 (i.e., identical

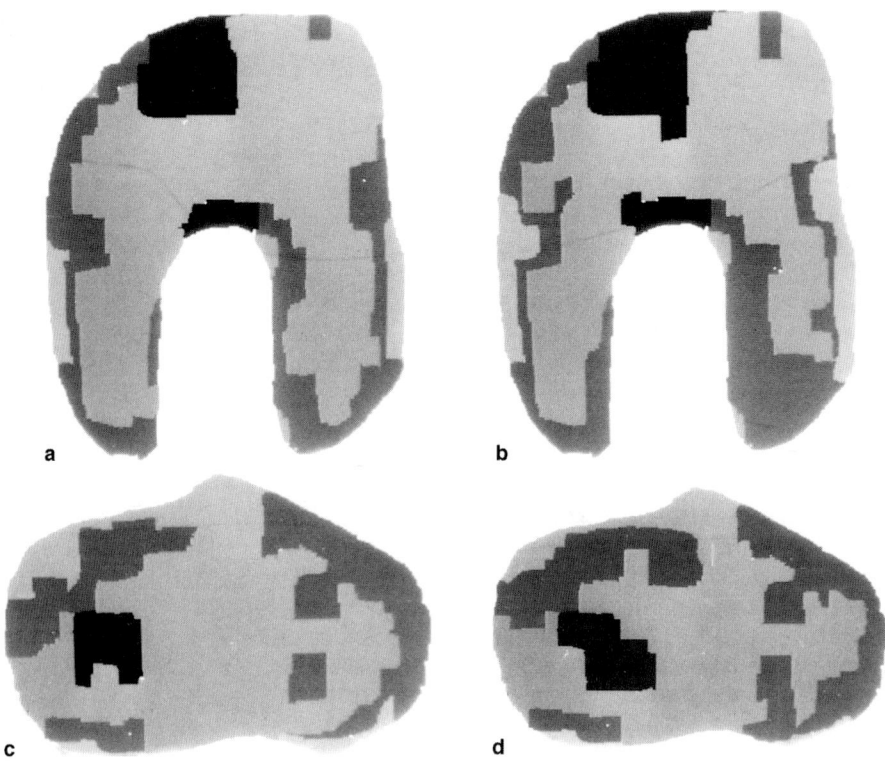

Fig. 11a-d. Density maps of the subchondral mineralization (a) in the femoral condyles (inferoventral aspect) from primary sections, (b) in the same femoral condyles from secondary sections, (c) in the tibial plateau (cranial aspect) from primary sections and (d) in the same tibial plateau from secondary sections

Table 9. Correlation coefficients and deviation from the straight slope (= 1) for the specimens examined

Specimen	Correlation coefficient	Deviation from the straight slope (= 1) (%)
1	0.984	− 3.2
2	0.995	− 6.2
3	0.994	− 3.1
4	0.987	−10.0
5	0.992	− 6.5
6	0.987	− 0.2
7	0.983	− 7.5
8	0.989	− 6.7
9	0.985	− 8.3
10	0.992	−01.3
11	0.996	− 0.2

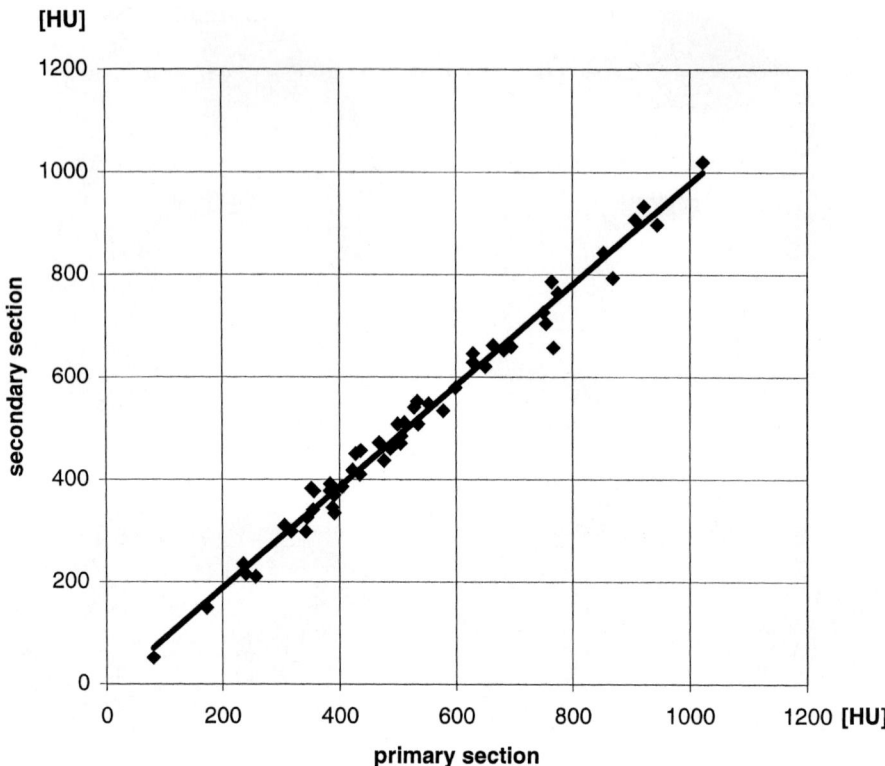

Fig. 12. Regression line of the *HU* values in corresponding sections of a knee joint (*x-axis*: *HU* values from primary sections, *y-axis*: *HU* values from secondary sections)

density values in both images) lies between +0.2% and –10%, depending on the degree of agreement in each case.

A further step in the quantitative determination consists of measuring the extension of the area of the single density intervals (200–400 HU, 400–600 HU, 600–800 HU, 1000 HU) in corresponding primary and secondary images (Fig. 13). The agreement lay between 62% and 86%, with a mean value of 82% (Table 10).

Because of the good agreement between the results produced by the primary and secondary sections, the latter can also be used for CT OAM. This means that CT OAM is available for all joints and that, with the use of orthogonal sections, only the same errors that are encountered with primary sections need to be taken into account.

Because of the practical difficulties of the procedure and also in order to reduce the radiation, it is desirable, with some joints, to prepare transverse datasets so that a part of the articular surface is not perpendicular to the beam. We, therefore, undertook a further comparative investigation on the knee joint in order to see whether thin transverse sections (only 1 mm) would lead to the same results as those obtained from 2-mm sagittal sections, which meet all parts of the articular surface of the knee joint

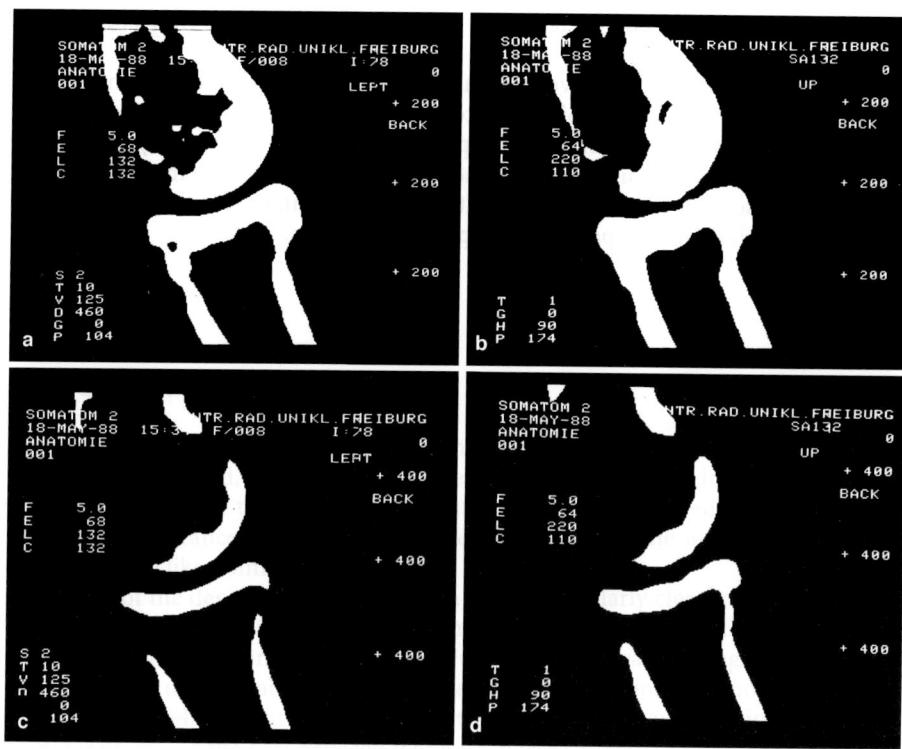

Fig. 13a-d. Comparison of the extent of the area of the different density zones in sagittal sections of the elbow joint. *White* areas are regions in which the density values are higher than the value indicated on the right side. **a.** Areas higher than 200 HU in the primary section. **b.** Areas higher than 200 HU in the secondary section. **c.** Areas higher than 400 HU in the primary section. **d.** Areas higher than 400 HU in the secondary section

Table 10. Number of imaging points of the same coordinates which show the same density in the primary and secondary sections (degree of agreement expressed in percentages).

Pair 1	71 %	Pair 9	77 %
Pair 2	84 %	Pair 10	79 %
Pair 3	78 %	Pair 11	70 %
Pair 4	67 %	Pair 12	82 %
Pair 5	81 %	Pair 13	77 %
Pair 6	80 %	Pair 14	73 %
Pair 7	73 %	Pair 15	62 %
Pair 8	86 %	–	–

at right angles to the beam. Comparison of the reconstructed pattern of mineralization from the surfaces of identical knees, in one instance from datasets of 1-mm transverse sections and in the other from those of 2-mm sagittal sections, revealed a high level of agreement with regard to the position of the maxima, although there are slight differences in the absolute values. These differences are primarily attributable to the

5.2
Shoulder Joint

5.2.1
Control Group

The articular surface of the glenoid cavity was subdivided for recording the position of each observed maximum. This resulted in three types (Fig. 15), which are shown in Table 11. The correlation of the occurrence of the different types with age and sex is given in tables 12 and 13.

Table 11. Classification of the different density types in the glenoid cavity found in the control group

Type I	A ventral maximum
Type II	Ventral and dorsal maxima, with a significantly less mineralized region lying between them
Type III	A central maximum

Table 12. Frequency of each type in relationship to the age of the subject

Type	16 years ($n = 25$)	26 years ($n = 20$)	36 years ($n = 16$)	45 years ($n = 9$)
I	32%	35%	47.0%	11.1%
II	44%	50%	46.8%	55.6%
III	24%	15%	6.3%	33.3%

Table 13. Frequency of each type in relationship to the sex of the subject

Type	Female ($n = 13$)	Male ($n = 57$)
I	46.2%	32.4%
II	38.4%	50.1%
III	15.4%	17.5%

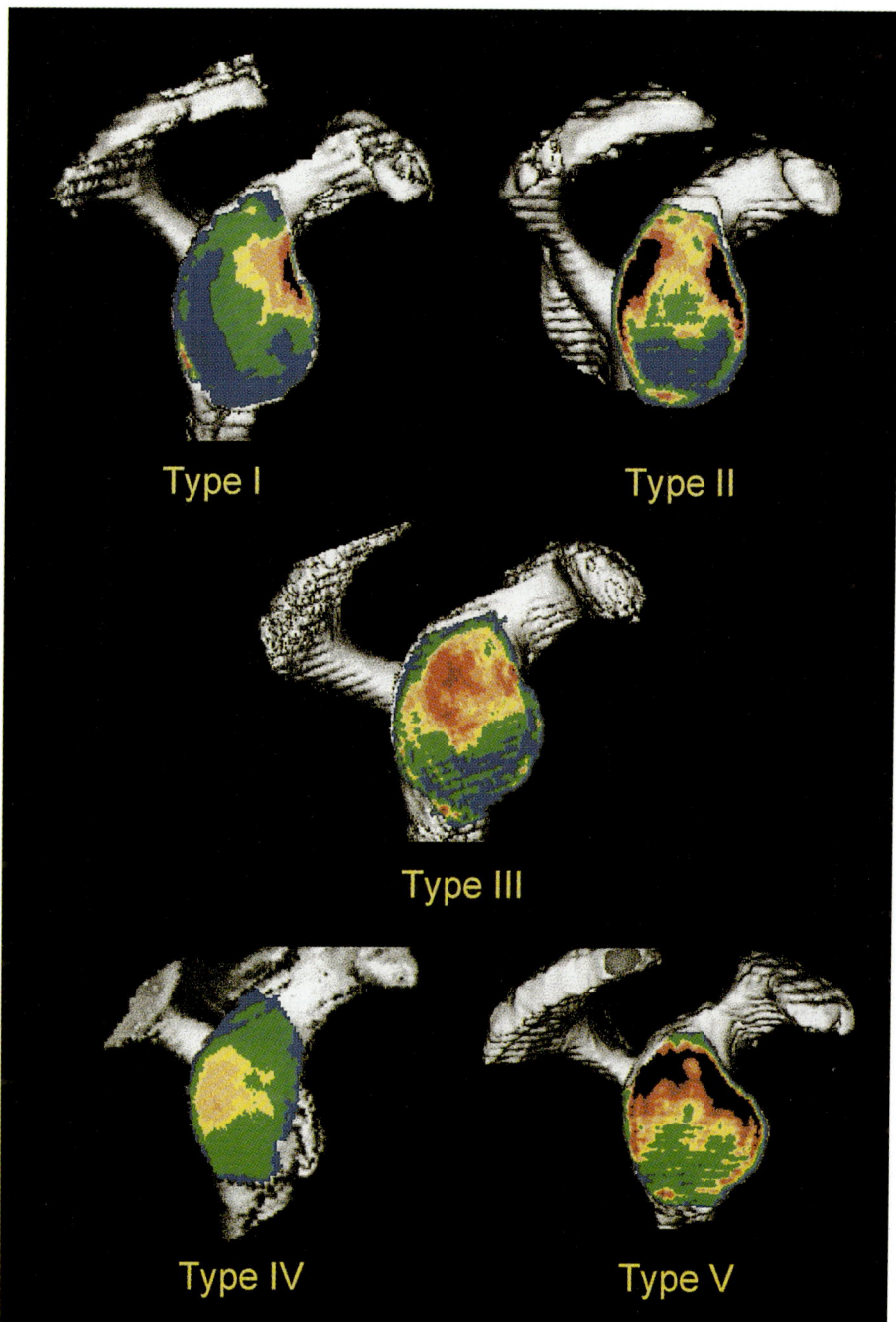

Type I

Type II

Type III

Type IV

Type V

Fig. 15a-e. Density maps of the glenoid cavity (lateral aspect). **a.** *Type I* (a ventral maximum). **b.** *Type II* (ventral and dorsal maxima). **c.** *Type III* (a central maximum). **d.** *Type IV* (a dorsal maximum). **e.** *Type V* (a cranial maximum)

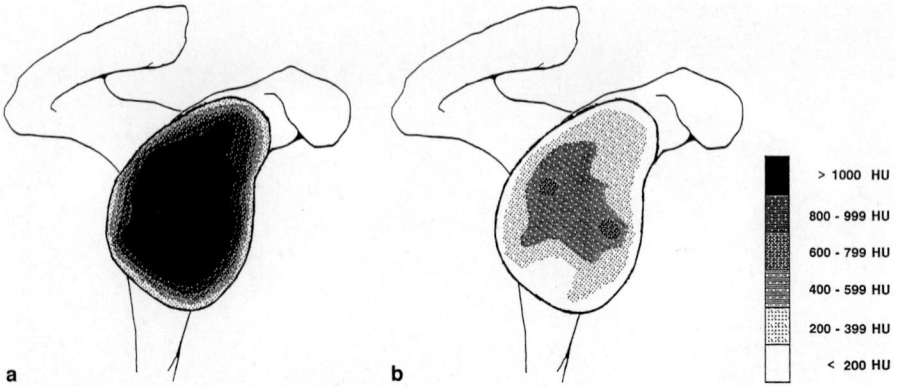

Fig. 16a, b. Density maps of the glenoid cavity (lateral aspect) (**a**) from a male gymnast and (**b**) from a patient with an unreduced traumatic dislocation of the shoulder joint

5.2.2
Mineralization Patterns of Athletes

The total mineralization of all gymnasts (Fig. 16a) was greater than that of the population as a whole. Two thirds of them revealed an extended central localization of the zone of greatest density (type III) and, in approximately one third, there was a lateral displacement of the maximum dorsally.

5.2.3
Mineralization Patterns in Cases of Unreduced Traumatic Dislocation of the Shoulder Joint

Two patients, in whom the humeral head remained outside the joint socket for an extended period of time (one for 1.5 months and the other for 3 months), showed bicentric patterns of type II, with the total mineralization distinctly less than in normal subjects (Fig. 16b).

5.3
Elbow Joint

In most cases (Fig. 17), the distal articular facets (the trochlear notch and fovea of the radial head) were more highly mineralized than their proximal counterparts (the trochlea and the capitulum). In each case, the mineralization within the joint surfaces showed a typical pattern of distribution. A central density maximum was almost invariably to be found in the fovea of the radial head, with the mineralization falling off concentrically toward the margins; very rarely were two maxima seen. In the capitulum, the central region was most highly mineralized, although here it could be displaced toward the groove between the capitulum and the trochlea. Maximum

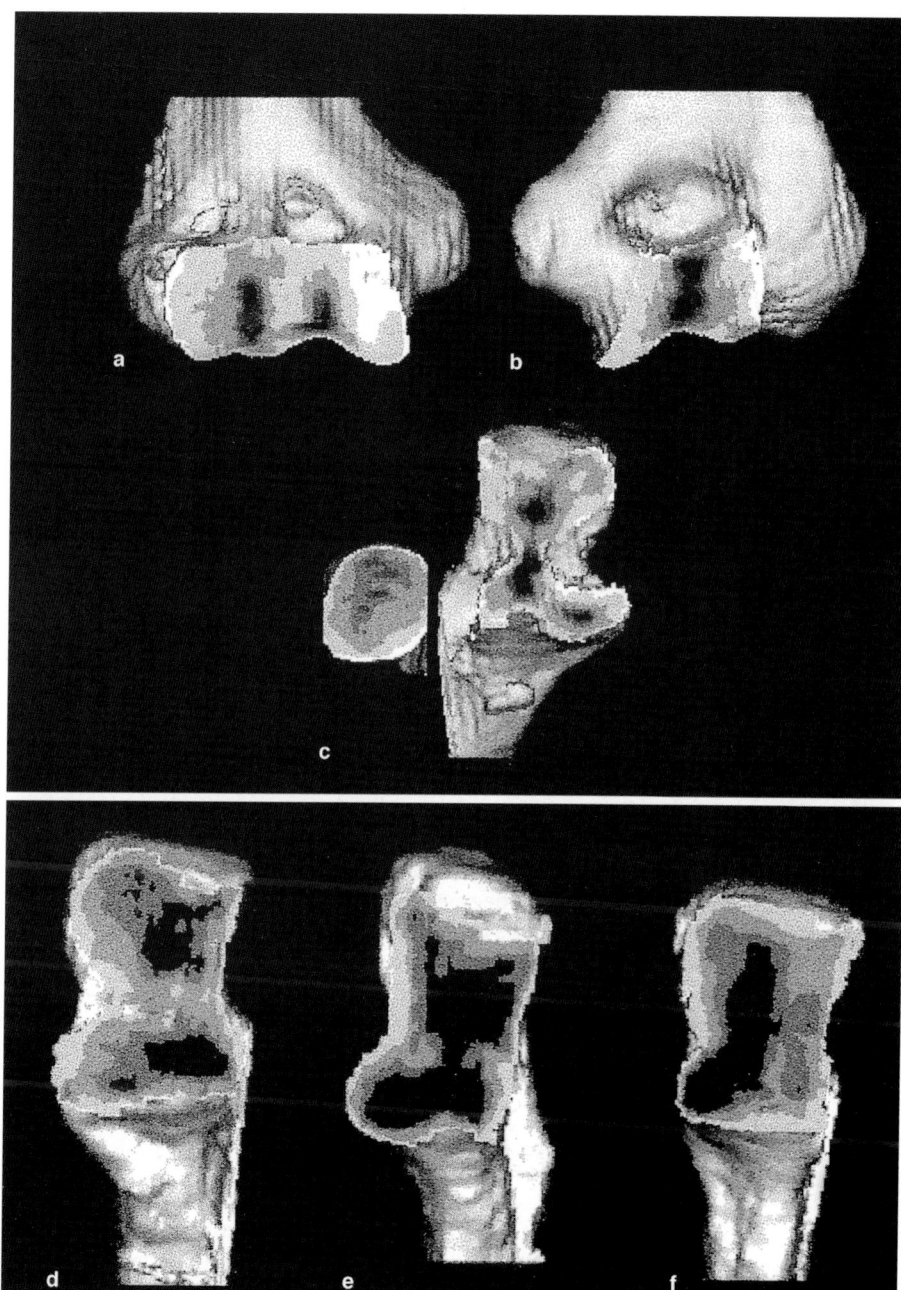

Fig. 17a-f. Density maps of the elbow joint: (**a**) humerus (ventral aspect), (**b**) humerus (dorsal aspect), (**c**) radius und ulna (ventrosuperior aspect), (**d**) ulna (superior aspect) with completely divided articular surface, (**e**). ulna (superior aspect) with continuous articular surface and (**f**) ulna (superior aspect) with medially divided surface

mineralization was found in the central and dorsal parts of the joint surface in the neighborhood of the above-mentioned groove.

Unlike the radius, the trochlear notch usually revealed a bicentric distribution pattern of subchondral density, one maximum being found in the dorsal region of the joint surface and one in the ventral region. The latter frequently extended right out to the medial edge of the surface. Mineralization in the depth of the trochlear notch was, on the other hand, considerably reduced.

When the distribution pattern was examined in terms of the joint morphology (varying forms of the cartilaginous covering), some differences were found among the three groups. In the group with the completely divided surface (Fig. 17d), a prominent bicentric pattern could be seen, mineralization in the depths of the notch being on average some 300 HU lower than beneath the ventral and dorsal parts of the joint surface. In the second group (medial side divided and lateral side covered by a continuous layer of cartilage (Fig. 17f), ventral and dorsal maxima were also found, although mineralization in the depths of the notch was, on average, only reduced by 100 HU relative to the two maxima. In the group with a continuous cartilaginous covering both medially and laterally (Fig. 17e), one also frequently encountered a bicentric distribution. However, occasionally a monocentric pattern with a single maximum in the central or dorsal region of the joint surface could be found. Mineralization in the depth of the notch, however, clearly was more pronounced in relation to the ventral and dorsal aspects of the articular surface than in the two other groups.

5.4
Radiocarpal Joint

In normal cases, the distal joint surface of the radius presented two density maxima. In 7 out of 17 hands, the maximum closer to the ulna had a relatively higher density (Fig. 18b). In seven hands, the radially located maximum showed a higher (Fig. 18a) and in three hands, showed the same degree of mineralization (Fig. 18c).

A comparison of the maxima on the opposed articular surfaces of the carpal bones revealed a higher degree of mineralization in the scaphoid than in the lunate in 13 hands.

Comparison of both sides in right-handed people revealed, in most cases, a greater degree of mineralization on the right than on the left.

5.5
Hip Joint

In all cases, the maximal subchondral bone density was found at the periphery of the acetabular cup corresponding to the limbus. Within the articular surface, zones of maximum bone density were found in the dorsal and ventral regions of the dome of the acetabulum in 14 of the 17 subjects under 60 years of age (Fig. 19a). In contrast to this, in 7 of 10 subjects of more than 60 years of age (Fig. 19b), the maximum density appeared in the central region (Table 14). One 65-year-old woman with marked generalized osteoporosis showed the greatest density in the center of the acetabular dome, together with a considerable reduction in the overall degree of mineralization.

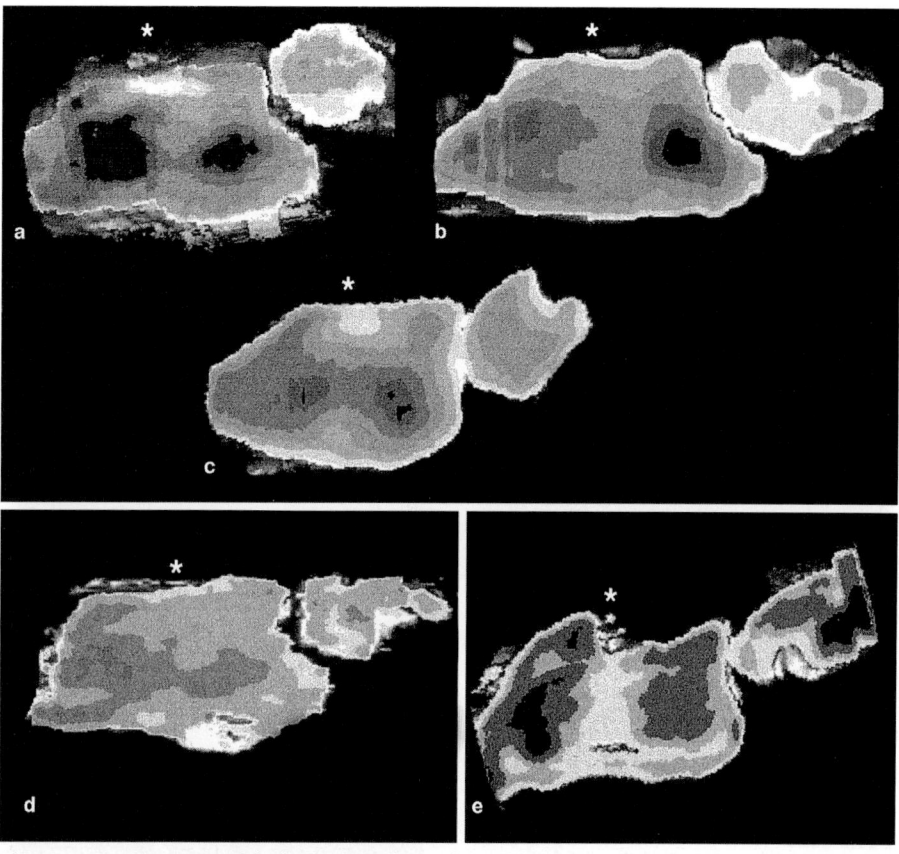

Fig. 18a-e. Density maps of the lower articular surface of the radius (distal aspect, * dorsal): (a) type I (higher maximum in the scaphoid fossa), (b) type II (higher maximum in the lunate fossa), (c) type III (equally balanced maxima), (d) from a patient with Kienböck's disease and (e) from a patient with a badly reduced distal radius fracture

Table 14. Position of the subchondral density maxima in the acetabulum in relationship to age ($n = 20$)

	Central	Ventral and dorsal
18–59 years	1	11
60–90 years	6	2

A small maximum was found in the posterior horn and an additional broader area of high density was recorded in the anterior horn.

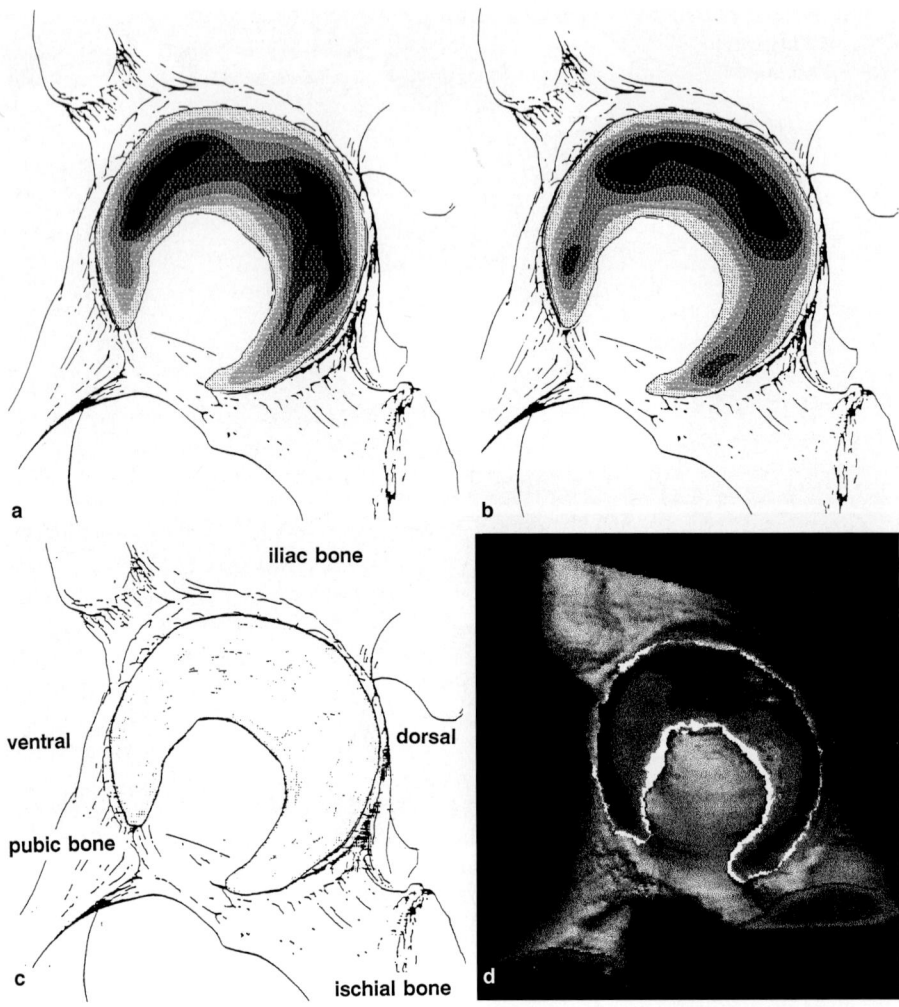

iliac bone

ventral   dorsal

pubic bone

ischial bone

Fig. 19a-d. Density maps of the lunate facet of a left hip joint (lateral aspect) (**a**) of a healthy 18-year-old girl, (**b**) of a healthy 68-year-old woman and (**d**) of a 34-year-old patient with hip dysplasia

5.6
Femorotibial Joint

In normal subjects (Fig. 20a), there was a central density maximum in the medial and lateral compartments of the tibial plateau. From each of these, the density values decreased concentrically toward the periphery. On average, the density of the maximum in the medial compartment was about 200 HU greater than that of the lateral.

Fig. 20a-d. Density maps of subchondral mineralization (+ lateral) (**a**) of the tibial plateau (cranial aspect) of a healthy young person, (**b**) of the femoral condyles (ventral aspect) of a patient with genu varum, (**c**) of the tibial plateau (cranial aspect) of a patient with genu varum and (**d**) of the tibial plateau (cranial aspect) of a patient with genu valgum

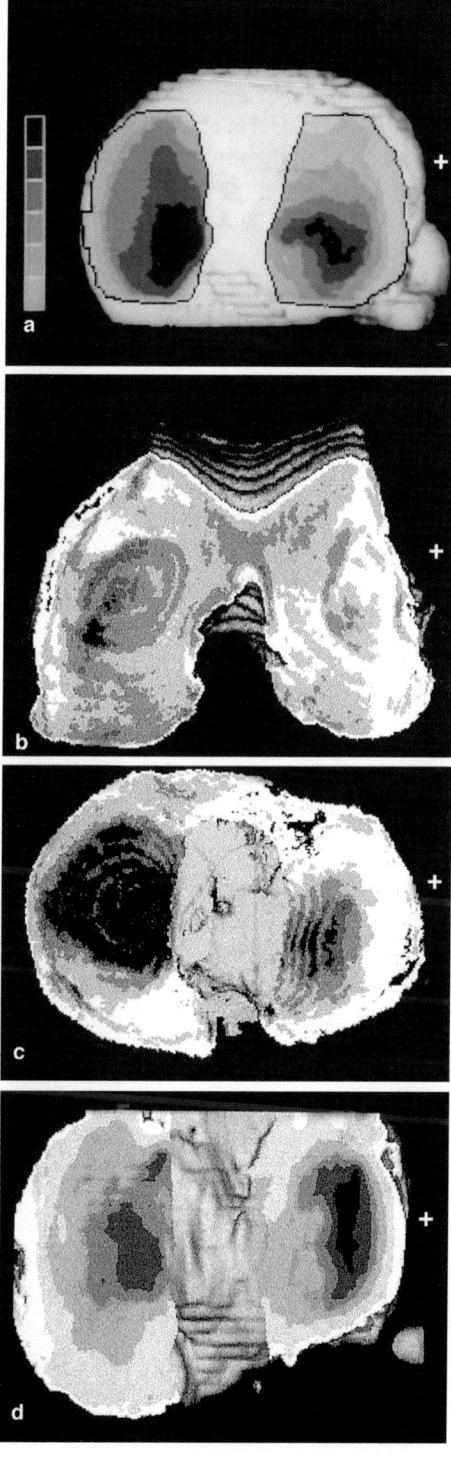

5.7
Femoropatellar Joint

The principal subchondral density maximum was constantly found in the proximal half of the lateral facet of the articular surface of the patella (Fig. 21). The density values decreased peripherally in a steep gradient in the proximolateral direction and more gradually in the opposite direction. In about a third of cases, there was a subsidiary maximum on the medial facet near the secondary ridge.

The density values for the corresponding joint surface of the femur (articular surface for the patella) were always about 150 HU lower.

5.8
Ankle Joint

In the upper surfaces of the talus, two different mineralization patterns are found. Type I is a bicentric pattern with both medial and lateral maxima in the joint surface (Fig. 22a). The greatest density was found along the medial ridge of the upper articular surface and was included in the medial surface; the lateral maximum on the upper part of the talus did not involve the lateral edge. The central region was distinctly less mineralized. There was a further maximum in the middle of the lateral surface.

Fig. 21. Density maps of the right femoropatellar joint of a healthy person (patella = dorsal aspect, corresponding aricular surface of the femur = ventral aspect)

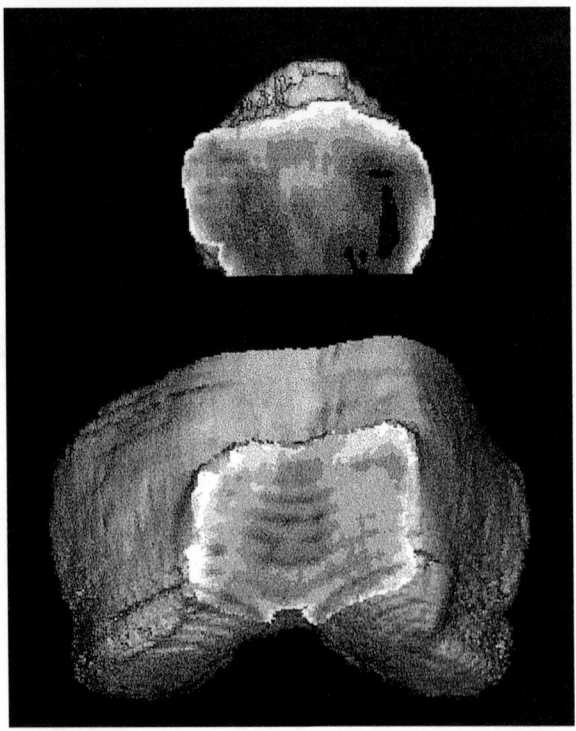

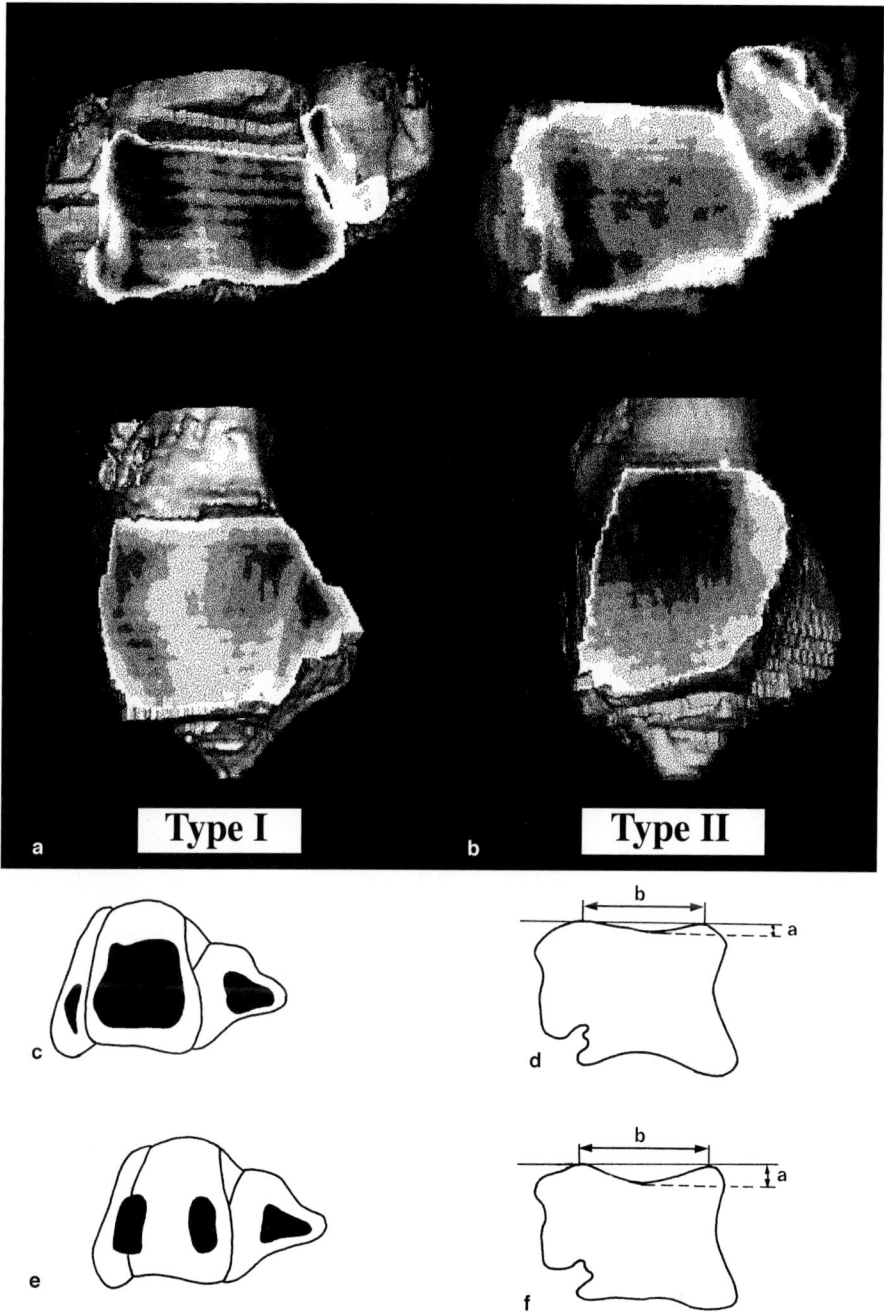

Fig. 22a-f. Density maps of the articular surfaces of a right ankle joint. **a.** Talar (cranial aspect) and lower tibial articular surface (distal aspect) of a young person. **b.** Talar (cranial aspect) and lower tibial articular surface (distal aspect) of an old person. **c.** Monocentric pattern of the upper talar surface (cranial aspect). **d.** Flat trochlear notch (frontal view). **e.** Bicentric pattern of the upper talar surface (cranial aspect). **f.** Deep trochlear notch (frontal view)

In contrast to this bicentric pattern, type II (Fig. 22b) usually presented only one maximum (monocentric pattern), which was an extension of the highest density stage over the whole surface and was localized centrally.

On the other hand, the corresponding surfaces of the malleoli presented a uniform distribution pattern. The distal part of the tibial articular surface always had its region of highest density in the transitional zone between its inferior articular surface and that of the medial malleolus. There was usually a lateral subsidiary maximum, which was, however, significantly less pronounced. Both maxima were preferentially located in the anterior region.

Measurement of the frontal talar profile (Riede et al. 1971) showed a deeper trochlear notch in young persons(Fig. 22d), becoming flatter with advancing age.

A correlation was found between the occurrence of the different density patterns and the depth of the trochlear notch; in persons with a deeper notch, the bicentric pattern was found, whereas the monocentric pattern occurred in persons with a flatter notch.

Comparison of the various types of mineralization with the talar profile quotient made it possible to associate a flat talus with the central monocentric pattern, whereas with a deeper leading groove, the pattern was bicentric.

6 Pathological Mineralization Patterns

6.1
Vertebral Column

6.1.1
Low-Grade Scoliosis

The findings in the vertebral joints in cases of scoliotic curvature (Fig. 14b) differed considerably, depending on whether they were associated with the convex or the concave side. On the concave side, the density was considerably reduced and the maximum values varied by as much as 500 HU. On the convex side, the density was greatly increased and the zones of highest density showed a marked surface extension, being predominantly medial in position.

6.2
Shoulder Joint

Unlike the joint sockets of healthy people, in which three different patterns of mineralization were found, in patients with diseases of the shoulder joint, five different types were recognized (Fig. 15).

6.2.1
Comparison Between Individuals with Healthy
and Diseased Shoulders

In 71 patients with a diagnosis of unilateral shoulder disease, the type observed was identical on both sides in 36.6% ($n = 27$) of cases. Type II was most frequently represented (59.3%), followed by type I (29.6%), type IV (7.4%) and type V (3.7%). Type III never appeared on both sides in one individual. In the 44 patients for whom the type on the diseased side differed from that on the healthy side, there was a large variety of combinations of distribution patterns.

Table 15. Classification of the different density types in the glenoid cavity found in the control group

Type I	A ventral maximum
Type II	A ventral and a dorsal maximum, the central region between them being considerably less mineralized
Type III	A central maximum
Type IV[a]	A dorsal maximum
Type V[a]	A cranial maximum

[a] Types not found in normal shoulders

Table 16. Frequency of the type found in relationship to the diagnosis

Diagnosis	Type I	Type II	Type III	Type IV	Type V
Ventral instability ($n = 62$)	24.2%	38.7%	12.9%	16.1%	8.0%
Ventroinferior instability ($n = 17$)	23.5%	35.3%	29.4%	5.9%	5.9%
Ventrodorsal instability ($n = 3$)	0%	33.3%	33.3%	33.3%	0%
Dorsal instability ($n = 7$)	14.3%	57.1%	14.3%	14.3%	0%
Multi-directional instability ($n = 5$)	40.0%	40.0%	0%	0%	20.0%
Rupture of the rotator cuff ($n = 5$)	0%	40.0%	0%	20.0%	40.0%
Osteoarthrosis ($n = 3$)	0%	33.3%	0%	33.3%	33.3%

6.3
Radiocarpal Joint

6.3.1
Healed Defective Repositioning of a Distal Fracture of the Radius

In the presence of a badly reduced distal fracture of the radius (Fig. 18e), the density pattern deviated from the normal and there was a displacement of the density maxima towards the dorsal part of the carpal surface.

6.3.2
Kienböck's Disease (Avascular Necrosis of the Lunate)

Nine of eleven patients with lunatomalacia (Fig. 18d) showed significantly reduced total mineralization in comparison with healthy subjects. This was also found on the unaffected side in six cases.

In eight patients, a density maximum was present in the radial facets for both the scaphoid and the lunate. In three patients with advanced lunatomalacia (stage III), there was no demonstrable density maximum in the facet for the lunate. In these patients, this maximum was also absent from the unaffected contralateral side.

6.4
Hip Joint

6.4.1
Dysplasia of the Hip

Apart from a small caudally located maximum in the posterior horn, the zone of highest density was recorded within the facies lunata in the central region of the roof of the socket in all of 10 patients with hip dysplasia (Fig. 19d). As already observed in normal subjects, a further zone of high density was seen in the region of the limbus, extending as a band along the edge of the socket.

6.5
Femorotibial Joint

6.5.1
Patterns in Patients With Malalignment of the Knee Joint

In cases of genu valgum (Fig. 20d), the density was considerably raised in the region of the lateral tibial condyle, where it showed an extended maximum. In the medial condyle, on the other hand, the density was reduced. With genu varum (Fig. 20c), the position was reversed: the density was significantly lowered laterally and markedly raised medially. Apart from an increase of the density in the medial condyle, the density maximum was displaced towards the medial edge. On the femoral side (Fig. 20b), the density in the medial condyle was significantly increased and in the lateral condyle, very much decreased.

6.5.2
Patterns Found in Patients with Genu Varum
After a Correction Osteotomy

An OAM examination 1 year after the operation showed significant changes in the density pattern compared with the initial situation. Apart from a more or less marked reduction in the total mineralization, the parameters "displacement of the position" and "alteration in density" of the density maxima made it possible to distinguish four groups (Fig. 23). In seven patients (group I), the medial maximum at the periphery had become displaced towards the center (in three cases by one sector and in four cases by two). In one case (group IV), the displacement was reversed, the maximum being further displaced medially towards the periphery. In six patients, there was, at first glance, no displacement. In three cases (group II), however, there was a reduction, and in three others (group III), an increase in the density in comparison with the preoperative situation, even though the surface maxima had not shifted.

If one compares the pre- and postoperative degree of mineralization for the lateral compartments separately, it appears that in group I there has been, in addition to the displacement, a reduction in the total mineralization, although laterally there was a significantly greater increase than that in groups III and IV. In group III, a significantly higher degree of mineralization developed bilaterally after the operation, but without any displacement. A comparison of the initial mineralization in all patients shows that, right from the start, there is a higher total mineralization in group II than in the other groups.

3-D evaluation of the total dataset showed that, when the preoperative and postoperative values were compared, the total number of density values in the region between 200 HU and 1000 HU was distinctly reduced after the operation (a preoperative value

Fig. 23. Density maps of patients with genu varum who had undergone a correction osteotomy. The corresponding *pre-* and *postoperative* patterns of a patient are arranged in rows

preoperative

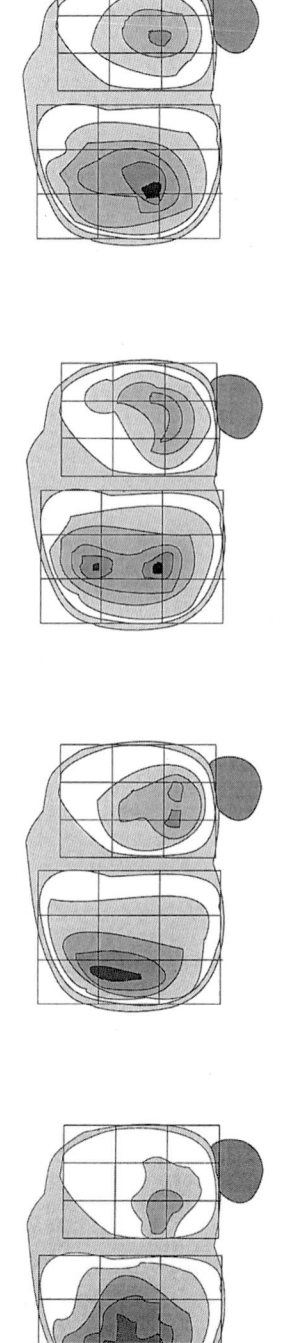

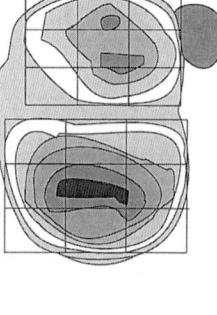

group 4

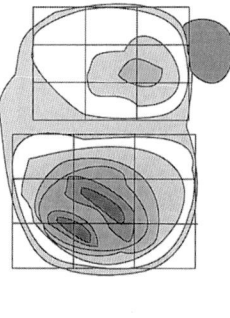

group 3

postoperative

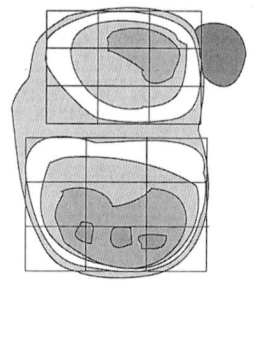

group 2

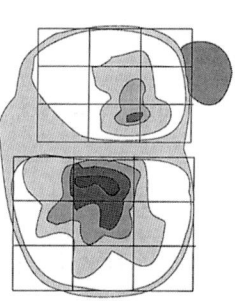

group 1

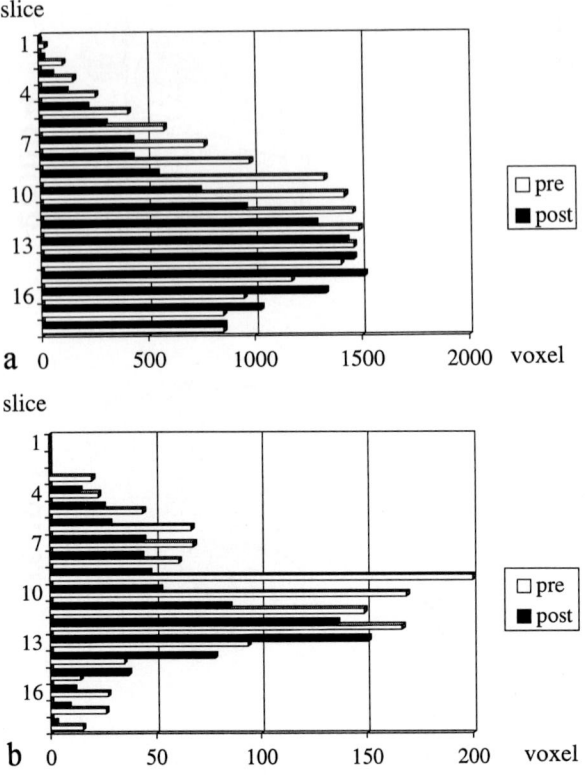

Fig. 24a, b. *Pre-* and *postoperative* comparison of the number of voxels per slice (**a**) for the whole density range (200–1000 HU) and (**b**) for the density range 501–600 HU

of 15725 compared with 12740, postoperatively). It could be shown, by means of the Wilcoxon test, that the decrease in the density values was highly significant within the 300 HU to 800 HU range in the first ten layers (p <N0.01), particularly between 500 HU and 600 HU. In the deeper layers, the significance was lower (p <N0.05) and from 800 HU upwards, it was not demonstrable.

The maximal values for the density steps involved were found preoperatively in layers 9–15, but after the operation they appeared in layers 12–17. This means that the maxima appeared at a deeper level after the operation and were, on average, 3.5 layers more distal.

A direct comparison of the pre- and postoperative totals for the density values of each single section made it clear that the preoperative tibial plateau showed a steeper and earlier increase in the total density values, so that its maxima were encountered in more cranially placed layers. The postoperative relationship revealed an increase that has been displaced distally and the greatest number of density values first appeared in the deeper layers (Fig. 24a). This was particularly striking in the density step 501 HU to 600 HU (Fig. 24b): the density region which was, in general. most strongly affected by the changes.

82

6.5.3
Patterns Following Reconstruction of the Anterior Cruciate Ligament With a Patellar Tendon Transplant (Sheep)

On the right (unoperated) control side (Fig. 25a), the subchondral mineralization maxima of the lateral joint surface lay centrally at the transition between the surfaces of the second and third order. In the medial joint surface, the maxima were also constantly found in the same transitional zone, but somewhat more ventrally placed than on the lateral side. The superimposition of all the maxima of group I (Fig. 26a) showed very clearly the unchanging form of the distribution pattern.

On the operated left side [48 weeks after open plastic surgery to the patellar tendon, (Fig. 25b)], the thickness maxima in the lateral joint surfaces lay close to the center of the surface of the second order, although in comparison with that in the control group, it was displaced somewhat in an anterior or posterior direction.

The density maxima in the medial joint surfaces lay in the dorsal region of the surface of the third order. In comparison with the control side, one could recognize a marked displacement of the medial density maxima dorsomedially and there was also an irregular dispersal of the surface density. Superimposition of the density maxima of all the animals examined (Fig. 26b) illustrated a constant distribution pattern after replacement of the anterior cruciate ligament, as already seen in the control group.

Statistical evaluation of the displacement of the density maxima on the operated side showed, in comparison with the control side (Fig. 27a, b), that the medial maximum had shifted further medially by about 2.5 IU ($p <$N0.05) and 2.2 IU ($p <$N0.05) dorsally. On the lateral side, the maximum was displaced by 1.5 IU ($p <$N0.05) laterally. There was no posteroanterior displacement.

6.5.4
Patterns After Medial Meniscectomy (Sheep)

Following meniscectomy (Fig. 25c, d), all the animals displayed a displacement of the density maxima in the medial tibial condyle towards the periphery. Statistical evaluation (Fig. 27e, f) showed, in comparison with the control side, a significant displacement in the medial joint surface, on average about 1.2 IU (ca. 2 mm) dorsally ($p <$N0.05) and medially ($p <$N0.01). In the lateral joint surface, on the other hand, there was no statistically demonstrable displacement of the maxima. Even so, they were invariably not, in comparison with the normal side, central.

6.5.5
Patterns After Primary Replacement of the Meniscus by an Autogenous Graft of the Patellar Tendon Covered With a Layer of Fascia Lata

Following replacement of the meniscus by an autogenous graft of the patellar tendon covered with a layer of fascia lata (Fig. 25e, f), the medial density maximum was displaced, in a dorsal direction only, by about 1.2 IU ($p <$N0.05). On the lateral side, there was again no displacement of the density maximum (Fig. 27c, d).

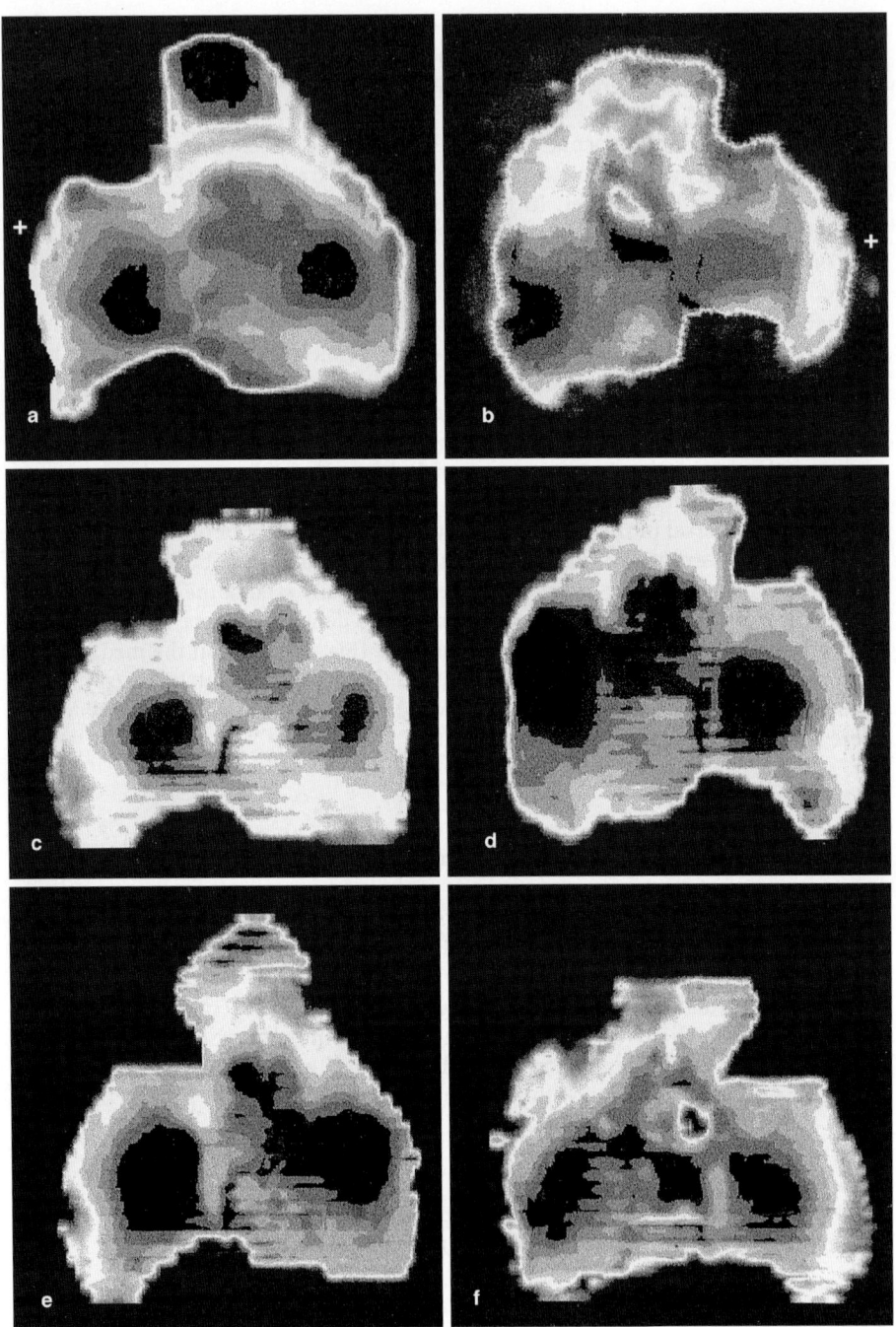

Fig. 25a–f. Density maps of the tibial plateau (cranial aspect, + lateral) (1) in sheep with ACL reconstruction: (**a**) control group and (**b**) 1 year after ACL reconstruction; (2) in sheep with medial meniscectomy: (**c**) control group and (**d**) 1 year after meniscectomy; and (3) in sheep with a primary meniscus replacement: (**e**) control group and (**f**) 1 year after primary meniscus replacement

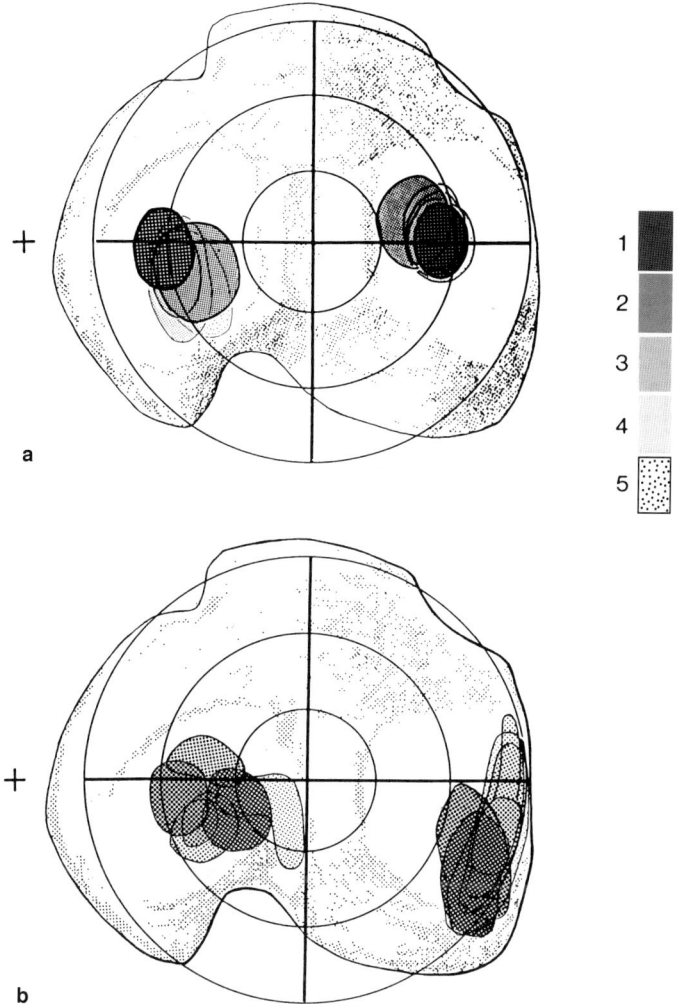

Fig. 26a, b. Superimposition of the density maxima of all animals investigated (cranial aspect, + lateral). The different gray values refer to the single specimens (n = 5). **a.** Control group. **b.** Operated group with ACL reconstruction

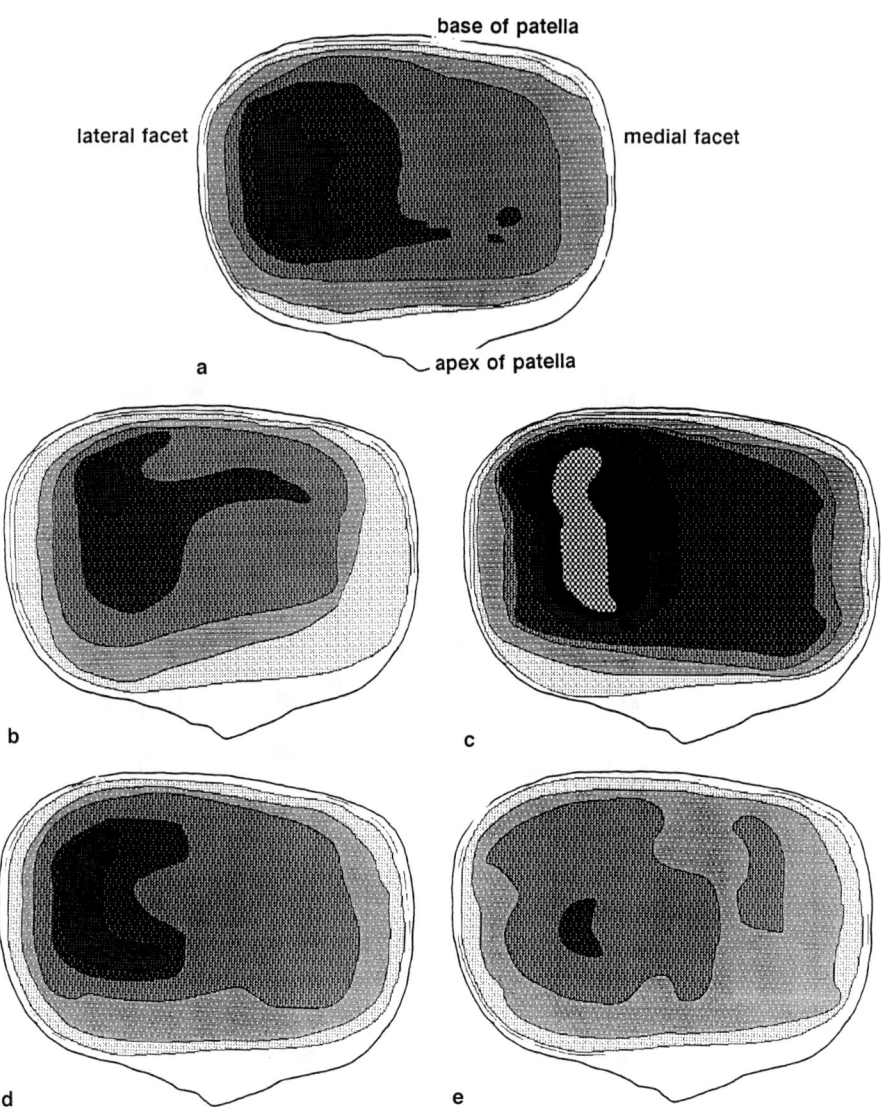

Fig. 28a–e. Density maps of the patella (dorsal aspect) in patients with (a) no cartilage damage, (b) laterally localized cartilage damage, grade I, (c) laterally localized cartilage damage, grade II, (d) medially localized cartilage damage, grade I and (e) medially localized cartilage damage, grade II

6.6.
Femoropatellar Joint

6.6.1
Patients with Retropatellar Pain

According to the classification of Outerbridge (1961), the arthroscopic findings could be summarized in the following way: six patients exhibited no recognizable cartilage damage; four patients showed cartilage damage on the medial side (grades I and II); eight patients had central cartilage damage (grades I and II); and in two patients, the cartilage damage (grades I and II) was located laterally.

In all patients, the main subchondral density maximum was found in the proximal part of the lateral patellar facet and in the central parts of the femoral joint surface. In patients without cartilage damage (Fig. 28a) and those with central damage, its position did not differ from the normal. In patients with cartilage damage laterally or medially, the position of the maxima was the same. Nevertheless, in those with damage on the medial side (Fig. 28d, e), the total mineralization in both joint surfaces was lowered and in those with lateral damage (Fig. 28b, c), raised.

The greater the degree of cartilage damage, the more marked was the decrease or increase of the total mineralization.

7 Factors Influencing The Development of Normal Patterns of Mineralization

The OAM findings obtained from different joints and among different samples of the population have clearly demonstrated that, within a particular joint and allowing for individual variation, regular and reproducible patterns of subchondral mineralization do indeed exist. It is beyond question that general factors, such as endocrines and alterations in metabolism, have an overall influence on the composition of bone. Osteoporosis, for instance, leads to a reduction in mineralization and certain pathological conditions such as Paget's disease bring about an increase. Nevertheless, the locally predominating mechanical conditions are probably decisive for the existence of local differences of subchondral bone density below the joint surface, even in the presence of the diseases already mentioned. Meanwhile the causal relationship between the long-term local stress and the subsequent subchondral mineralization seems to have been so clearly established that, from the distribution of the latter, conclusions may be drawn about the mechanical conditions within the joint.

The degree and distribution of the subchondral mineralization below the articular surface, therefore, reflect to some degree the stress acting on it (Fig. 29). To be more precise, the distribution of stress is determined by the magnitude and the penetration point of the resultant force, by the size and position of the contact surfaces and by the shape of the other joint component. These factors themselves are again very much dependent on the forces exerted by the periarticular muscles and ligaments.

From the normal pattern of mineralization, the different factors influencing the distribution of stress can now be described. It must, however, be emphasized that the mineralization pattern does not allow conclusions to be drawn about one particular factor. Caution is also necessary when directly comparing the results of studies of pressure distribution or the determination of contact surfaces. These are generally concerned with instantaneous recordings of selected loading conditions, the outcome of which is largely determined by the experimental design. Since these experiments can normally only be carried out on postmortem specimens, the influence of the musculature and neuromuscular control cannot be satisfactorily taken into account, because far too little is known about the individual interaction and fine coordination of the periarticular muscles. Mathematical models are also limited in the information they can provide, since what comes out of these models depends very much on what is put in. The mineralization pattern, on the other hand, deals with the biological reality of the long-term mechanics of the joint, although, it must be admitted, only as the integral of all the underlying factors.

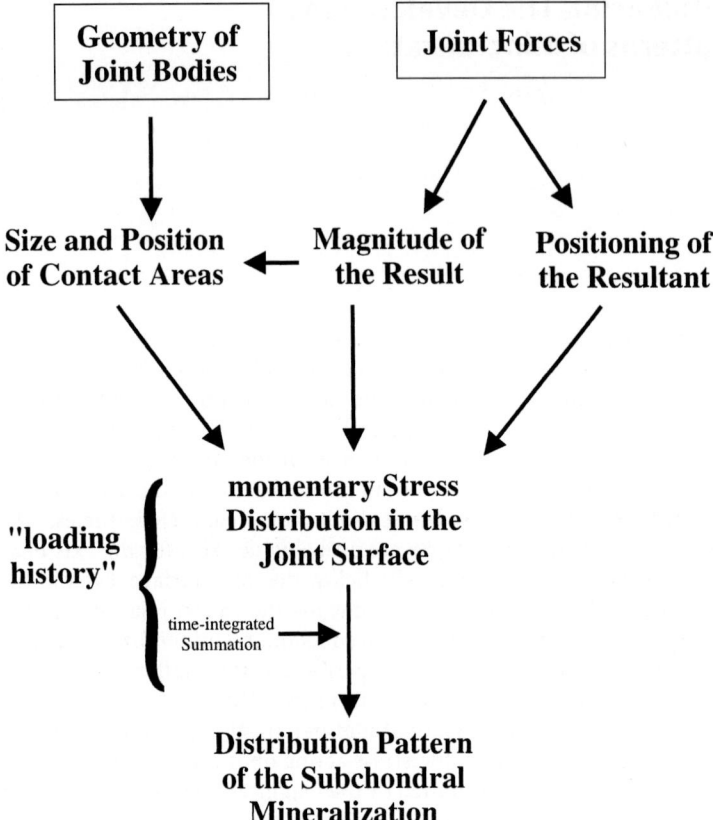

Fig. 29. Factors influencing the stress distribution within a joint surface

7.1
The Shape of the Joint Surfaces

7.1.1
Spherical and Hinge Joints

The fact that the distribution pattern is mechanically relevant is particularly obvious when one compares our work on the lunate surface with the results and interpretations of other authors. Although it seems surprising, at first glance, results obtained from young people have revealed that the zone of highest bone density is not central, but lies in the ventral and dorsal regions of the roof. It is surprising because, on the basis of the usual models of hip-joint mechanics regarded in the frontal plane, the region of greatest stress is seen at the highest point of the socket and one would anticipate that the region of greatest mineralization must lie here. This is the region that Pauwels

(1973) - admittedly, only in AP radiographs – put forward. However, an explanation for the apparent contradiction was provided by the investigations of Bullough et al. (1968), who observed that, in the unloaded joints of young subjects, the joint components are aspherical. This asphericity diminishes with increasing age. Additional study of the contact surfaces in the hip joint (Bullough et al. 1973) confirmed that, with slight loading in young people, the contact surfaces lie in the anterior and posterior regions of the acetabulum and that this is independent of the relative position of the bones.

In contrast, in older subjects, the contact zone is often found in the roof of the socket. In their experiments on the position and extent of the contact surfaces, Greenwald and Haynes (1972) observed no contact zone here in 83% of subjects under the age of 60 years, but rather in only 36% of those over 60 years. These figures agree with our morphological results (see Table 14).

According to the studies of Miyanga et al. (1984), who examined loading at the hip joint in the primarily incongruous joints, the contact areas merge in the acetabular roof at loads of about 60% of the body weight, with a corresponding increase in the contact surface and lower compressive stress. This is because, as further experiments on the deformation of the cartilage and subchondral bone under loading have demonstrated, the hip joint as a whole possesses a certain viscoelasticity (Miyanaga et al. 1984). Recent quantitative investigations into the distribution of the width of the anatomical joint space using polyether casts under different degrees of loading and investigations into the distribution of pressure with Fuji Prescale film (Eisenhart-Rothe et al. 1996, 1997) have confirmed the existence of different degrees of incongruity in different individuals. Furthermore, it was not possible to find any dependence of the incongruity of the joint on the age of the subject. The small number of specimens examined ($n = 12$) must, however, be taken into account.

According to Bullough (1981), the biological advantages of this incongruity are twofold. Because of the deformation of the cartilage and underlying bone, an increase in the loading leads to a comparatively small increase in the compressive stress, so that this type of joint is not only very stable, but is also best able to distribute and reduce the load. In addition to this, because of the incongruity present in the unloaded or only slightly loaded joint, the relatively high and locally mobile pressure gradient facilitates fluid exchange between the synovial fluid and articular cartilage, thus ensuring that the latter receives adequate nourishment (Afoke et al. 1980). On the other hand, when there is complete permanent congruity of the joint surfaces, the absence of a space severely impairs the circulation of the synovial fluid and this restricts the metabolic activity of the chondrocytes.

It is, therefore, reasonable to suppose that the components of a normal hip joint are constructed in this way, so that they may not only provide good conditions for lubrication and nourishment of the normally loaded cartilage, but also ensure a stress-dependent (or dynamic) type of pressure transmission and distribution.

Whether this incongruity is to be attributed to the structural properties of the bone (Ogston 1878) or to the functional swelling of the cartilage, suggested by several other authors (Ingelmark and Ekholm 1948; Ekholm and Ingelmark 1952; Walker 1973; Oberländer 1978) cannot yet be decided.

A similar situation can be observed in the humero-ulnar joint, which consists of a joint socket (the trochlear notch) and a joint head (the trochlea). Individual differences, however, have established that this is related to the morphology of the cartilaginous covering of the proximal end of the ulna. Heine (1925) observed, during the examination of 500 elbow joints, that the trochlear notch always has a continuous cartilaginous surface during the first decade of life. In adults, however, there is complete separation of the joint surface by a central bony furrow in two thirds of subjects, a medially separated but laterally continuous surface in one third and, only very rarely, a bilateral completely continuous surface. These percentages have been confirmed by the work of Tillmann (1978b, 1971). These authors attributed the subdivision of the cartilaginous surface to a regression of the embryonic central articular cartilage. However, the reduction in cartilage thickness in the socket can be plausibly explained in terms of Pauwels' (1965) theory of causal histogenesis, since the supporting tissue reacts by regression to the reduction in contact stress caused by the incongruity. The bicentric density pattern which we found in the majority of adults supports this suggestion.

In the rare cases of a continuous articular surface, the central density maximum, observed in a few individuals, suggests a central transmission of pressure in these joints, so that a lesser degree of incongruity can at least be assumed. These conclusions are supported by the work of Goodfellow and Bullough (1967), who found different degrees of congruity in the humero-ulnar joint, depending on the age of the subject. The mixed type of joint (subdivided medially, covered by continuous cartilaginous layer laterally) suggests the existence of different degrees of incongruity in the medial and lateral compartments, although it is not clear to what these differences may be attributed.

Determination of the contact areas has shown that with less loading the pressure is transmitted through dorsal and ventral contact surfaces (Goodfellow and Bullough 1967; Walker 1977; Goel et al. 1982). As the load increases, the contact surfaces extend themselves centrally. Eckstein et al. (1993b), who determined the contact surfaces while taking into account the varying morphology of the ulna (see above), were able to show that in joints with a continuous cartilaginous covering the contact surfaces are more centrally localized under lower force. In comparison with those joints with subdivided surfaces, the contact areas become extended centrally under relatively low load and this may lead to an unfavorable distribution of the stress. The subchondral mineralization maxima, which indicate central stress maxima, support this hypothesis. For the groups of joints with a continuous cartilaginous covering, a lesser degree of incongruity must therefore be assumed.

Examining the results obtained from the hip joint and the humero-ulnar joint together allows one to conclude that this incongruity represents a general biological principle of construction which produces optimal conditions. The more favorable stress distribution under different degrees of loading have been numerically demonstrated in a finite element model by Eckstein et al. (1994a), who simulated the different stress distributions for varying degrees of incongruity.

A similar construction principle to that found in the hip and humero-ulnar joints also appears to be present in the shoulder joint. The density pattern shows – in the manner already described for the acetabulum – varying age-dependent patterns in normal subjects. However, unlike the situation in the joints already discussed, three different distribution patterns were found in the glenoid cavity in a normal sample which consisted predominantly of young people. In addition to a bicentric pattern with a ventral and a dorsal maximum, separated by a central region of lower mineralization (type II), one encounters an equal number of monocentric patterns with a ventrally placed maximum (type I). Whether type I or II is more common appears to depend on sex; whereas type II was found in about half of the men examined, in the women, type I predominated by about 46%. Less common, but appearing more frequently with increasing age, there is a maximum in the center of the socket (type III). The central localization of the maximum in many of the older subjects tends to be obscured by the degenerative changes often found in this position (Kohn et al. 1985).

A similar mechanical principle of joint construction to that in the hip joint can be assumed here. According to Saha (1973), who carried out spherometric investigations on the shoulders of cadavers, there are various types of shoulder joints which can be recognized from their differing degrees of incongruity. In his reports, however, one can find neither the age nor frequency distribution of the various types. Soslowsky et al. (1989), who measured the sphericity of the humeral head and glenoid cavity by means of a stereophotometric technique, found variations in the length of the radii in both the socket and in the corresponding humeral head. They stated that "this incongruency may be important in understanding fluid flow in the joint space and the resulting lubrication and load transmission processes". Therefore, the existence of an incongruity of the shoulder joint with ventral and dorsal contact areas in younger people may be suggested, since the central region of the articular cartilage of the glenoid cavity does not regularly serve as a contact surface as long as a particular value of the resultant is not exceeded. This provides the necessary conditions for the development of degenerative changes as the result of disuse. Owing to the reduction of the incongruity, this central region becomes a permanent contact area in older people, which is confirmed by the frequent degenerative changes seen here.

With gymnasts, however, where there is a widely extended area of greater mineralization, the initial situation is different. In some types of exercise, such as *Ringeturnen*, for instance, forces of up to five times the body weight can arise (Bodem et al. 1984). If, as a result of the often powerful forces acting, this region is continually under stress, it will obviously lead to a considerable increase in the mineralization here.

It is rather more difficult to explain the frequent appearance of type I (a monocentric ventral maximum) in normal subjects. The explanation may lie in the occurrence of the different types of socket that have already been described by Anetzberger and Putz (1996). A qualitative comparison of the various forms of the cavity with the accompanying position of the maximum makes it possible to assume that the drop-shaped form is associated with type II and the pear-shaped type (with its ventrally more extended component) with type I. One can also speculate that the ventral part of a pear-shaped cavity is subjected to greater bending stress, since it is less effectively supported by the spine of the scapula in comparison with the dorsal part.

7.1.2
Nonspherical Joints

As an example of a nonspherical type of joint, the subchondral mineralization patterns of the articular surfaces of the ankle joint were examined to obtain information about the mode of load transmission there. The density patterns observed were once again of two types. The commoner was a bicentric pattern with maxima in the area of the medial and lateral ridge of the trochlear surface of the talus. The inferior articular surface of the tibia constituted a mirror image of this. Less often there was a monocentric pattern with a central region of high density and this was found predominantly in older people.

The recording of individual differences in the form of the trochlea by means of a quantitative measurement of the varying depth of the leading sulcus of the trochlea in individual subjects (the frontal talus profile quotient of Riede 1971) has shown, in a comparative study of the various types of mineralization, that the flatter tali are associated with the central, monocentric patterns, whereas in the presence of a deeper leading sulcus the pattern is bicentric. In other words, a clear interdependence can be established between the extent of the mineralization and the shape of the individual joints.

The striking dependence of the talus profile quotient on age, seen even in our limited sample, is confirmed by the work of Riede et al. (1971) who was able to show from measurements made on more than 800 subjects that the depth of the trochlea is associated with a significant flattening of the leading sulcus as age increases. These results confirm that the exact surface relationship of the joint components, even for irregularly curved articular surfaces, such as those of the ankle joint, is a prerequisite for adequate tissue transmission of the stress. This supports Wynarsky and Green-wald's (1983) mathematical model, which postulates that primary incongruity ensures a more satisfactory stress distribution in adapting to varying loads.

The observation of a bicentric pressure transmission through peripheral supporting pillars is, therefore, not confined to spherical joints, but seems to be a general principle that is active in several of the larger joints of the human body.

This physiological incongruity is apparently not to be found in all joints. The appearance of central density maxima in the opposing surfaces of the humeroradial joint suggests a predominantly central transmission of the load. That this is bound up with the geometrical situation is suggested by the findings of Bünck (1990) and those of Eckstein et al . (1996)by MRI, who quantified the curvature and contact-surface relationships of this joint. Since the fovea of the radius has a flatter curvature than the capitulum of the humerus, it is inevitable that a central contact must exist, whatever the magnitude of the force applied may be.

It must be critically observed, at this point, that the existence of a physiological incongruity cannot be proved by the pattern of mineralization found by us. If, however, one compares these mineralization patterns with the results of various studies, whether they are experimental studies on cadaveric specimens or mathematical models, it is a plausible theory with biological advantages for all the structures taking part.

When interpreting the mineralization patterns, the possible deformation of the articular surface must also be taken into account. Since both the glenoid cavity and the acetabulum are integrated components of the shoulder girdle and pelvis, respectively, deformation of the whole structure could also contribute to the configuration of the pattern observed. This relationship was postulated a long time ago by Simkin et al. (1980), who reasoned that the greater thickness of the subchondral bone in the concave joint surface was due to a higher tensile stress acting there. This theory is supported by the results of a new finite element calculation on the humero-ulnar joint (Merz et al. 1997; Jacobs and Eckstein 1997). Their simulation of bone modelling in incongruous joints does indeed produce a bicentric density pattern, which nevertheless differs markedly from the calculated stress distribution in the joint surface. It does, however, agree well with the tangential tensile stresses which arise in the subchondral bone as the outer parts of the sockets are driven apart, but not with the contact stress. A parallel study by Eckstein et al. (1997) demonstrated the sagittal course of the split lines in the subchondral plate of the trochlear notch, which corresponded to the anticipated tensile stress in the model. Since the tensile stress in incongruous joints exceeds the pressure stress up to threefold, the authors conclude that, in spite of Pauwel's hypothesis, not only the compressive stress, but also the tensile stress determines the functional adaptation of the subchondral bone.

7.2
The Magnitude of the Joint Reaction Force

If the patterns found by us are relevant, it must follow that qualitative and quantitative changes take place during life. Since the loads acting on the joints can change during this period, whether due to an increase or decrease in activity or because of something which alters the mechanics of a joint (the defective repositioning of a healed fracture, for instance, or a correction osteotomy), the corresponding density pattern must show an altered picture for each change in conditions.

As is well known, continual remodelling of bone takes place as an adaptive reaction to each mechanical situation arising, and this also applies to the "bone remodelling" of a joint. Increased loading leads to an increase in density of the bone (Nilsson and Westlin 1971; Chamay and Tchantz 1972; Jones et al. 1977; Woo et al. 1981; Martin et al. 1980) and reduced loading to resorption (Issekutz et al. 1966; Donaldson et al. 1970; Whedon 1984).

That this also applies to subchondral bone is confirmed by the results obtained from gymnasts, who show density values significantly in excess of what is found in normal people. Results from ring gymnasts, in whom forces up to five times the normal body weight can arise at certain phases of the exercise (Bodem et al. 1984), make the higher mineralization comprehensible.

An example of reduced mineralization following adaptation to reduced stress was found in a patient with poliomyelitis, in whom the density maximum in the retropatellar joint on the affected side was in the same place, but showed a significantly lower degree of mineralization, even though the change here must have taken place over a very long time (Eckstein et al. 1993d). These results clearly show that the subchondral pattern of mineralization is, in terms of its total mineralization, dependent upon the magnitude of the joint reaction force. A long-term reduction in use leads to a significant reduction and increased loading to a reactive increase in the total mineralization.

The vertebral joints are segmental joints, inevitably working in combination. As joint components, the cartilage-covered surfaces of the articular processes of adjacent vertebrae come into contact with each other. They only partly represent simple geometrical shapes and even the transverse direction of the joint spaces, within single motion segments, cannot be grouped under a simple geometric principle in all regions of the column. The joint surfaces show a great variability, particularly in the lumbar column (Putz 1976). Here, it is necessary to distinguish between a more or less frontally oriented medial component and, extending outwards therefrom, a lateral sagittal component orientated dorsally or dorsolaterally, as has been described in detail by Lutz (1967) and Putz (1981). For these joints there is no general segmental center of curvature and the individual joint surfaces themselves are not regularly curved. Face-to-face contact between the joints occurs for most of the vertebral joints in particular positions which, in general, correspond to the normal bodily posture. A frequent loss of contact between the joint components even with slight movement is part of the normal function of these joints.

Indeed, the single CT sections from the OAM demonstration have already produced the surprising result that, in contrast to current belief, the zones of highest density are not located in the lateral component, but are found at the front of the medial component of the joint. This is surprising, because the lateral cancellous structure indicates a significant adaptation to bending stress (Putz 1985).

These morphological findings prove that the vertebral joints have an important role in pressure transmission through the vertebral column, in that they take up the accompanying shear stresses. As has been explained by Kummer (1981, 1991, 1992), amongst others, the joint reaction force in the vertebral column runs obliquely from above, downward and forward. Its course and, therewith, the sharing out of the proportional forces onto the discs and vertebral joints, depends on the prevailing bodily posture and on the position of the penetration point of the joint reaction force within the supporting surfaces of the motion segments. The nearer this point is to the edge, the more unilaterally is the stress-receiving part of the vertebra loaded. Since the resultant of the forces exerted by the body-weight and the pull of the deep back muscles is not perpendicular to the upper plates of the more caudal vertebral bodies, but is to a greater or lesser extent directed forward, it possesses a shearing component that tends to displace the upper vertebra anteriorly. This displacement is prevented by its inferior articular processes.

In this way, the anterior part of the joint is, except during extreme dorsiflexion, constantly subjected to a static compressive stress, and that accounts for the increased density in this region.

To what extent the individual vertebral joints are subjected to pressure depends among other things on the inclination of the articular surface to the long axis of the vertebral body. If these surfaces are parallel to the vertebral axis (i.e., at 90° to the upper plate), the ventral displacement component acts as a normal force. This is virtually the case in the lumbar region, which accounts for the increased stability of the column here.

The prevailing bodily posture likewise has great influence on the magnitude of the compressive stress in the vertebral joints. El-Bohy et al. (1989) was able to show, by directly measuring the pressure on the articular surfaces of the vertebral joints with a pressure transducer, that with an increasing flexion moment of the vertebral column caused by the powerful muscles which bring the body into postural equilibrium, there is a linear increase in the pressure acting through the vertebral joints. The greater the flexion, the greater the pressure forces in these joints.

Ueno and Liu's (1987) finite-element calculating model was further able to show that the vertebral joints, depending on the width of the joint space, must take over part of the load during the movement of rotation. They likewise found stress peak values in the medial facet during rotation. Investigations by Lin et al. (1978) on specimens also underlined the dependence of the degree of compressive stress in the vertebral joints on the prevailing bodily posture. As soon as shearing forces appear, which is normally the case during complex quasistatic movements, considerable force is transmitted through the vertebral joints.

The density maxima are found in the frontally orientated medial part of the joint surface, whereas the sagittally orientated lateral parts are significantly less mineralized and show an unambiguous adaptation to bending stress. If one brings these morpho-logical facts together with mechanical investigations on the vertebral column, it can be seen that the stresses on the medial and lateral parts are qualitatively different. Whereas the medial part is constantly subjected to high compressive stress and provides the most important limitation to flexion, the lateral part is particularly subjected to bending stresses and serves to limit rotation.

7.3
Position and Extent of the Contact Surfaces

The mineralization patterns of the articular facet of the patella determined by CT OAM are, so far as the size of the maxima and their position within the joint surface are concerned, in agreement with the X-ray densitometric findings of Tillmann and Brade (1980) and with our own results obtained from 30 anatomical specimens (Eckstein et al. 1992). The density maxima in the proximal part of the lateral patellar facet may be interpreted as an expression of the more frequent contact areas (Hehne 1983) of this surface under greater pressure (Maquet 1976) with the knee flexed between 90° and 140°. In the femoral patellar facet, the density maximum in the distal region of the joint corresponds, as in the patella, to the contact surface at an angle of flexion of 90°–120° (Hehne 1983). With increasing knee flexion, this region is subjected to a higher pressure stress than the proximal part of the joint (Maquet 1976) which, in this articular surface, again indicates a close relationship between subchondral mineralization and the principal stress.

The fact that, in all the cases examined, the density maxima of the patella exceeded those in the corresponding surface of the femur by about 150–300 HU, may be because the force introduced during flexion is distributed over a greater contact surface than that of the significantly smaller patella.

In contrast with the large joints of the human body, face-to-face contact of the components of most of the vertebral joints only arises in particular positions of the joint which frequently accompany a normal posture. Reduction in the contact surfaces, which often takes place with even slight movements, is a part of the normal function of these joints. Especially at the limit of a movement, the joint space widens from a peripheral point-like or linear contact to a wedge-shape contact (Dittmar 1930; Reich-mann et al. 1972). That means that gaping of the joint belongs to the normal pattern of movement of the vertebral column. Radiologically, this was demonstrated by Ravelli (1955) by the vacuum phenomenon which appears at the limit of a movement.

Because of their very small contact surfaces, gaping of the joint at the limit of a movement produces pressure peaks in the upper and lower parts of the joint, while the region between is, over a period of time, relatively unloaded. In the end, this loading pattern leads, because of the accompanying high static force in the peripheral parts, to increased mineralization of the upper and lower regions. Adams and Hutton (1983) concluded from this that in extreme flexion, when the shear force is greatest, contact is limited to the periphery and the resulting concentration of stress in the cartilage of the upper edge of the joint might possibly initiate degenerative changes. That would account for the degenerative changes in this region described by Schelbe (1987).

7.4
Penetration Point of the Joint Reaction Force

The point through which a resultant force is transmitted has a decisive influence on the distribution of stress within the articular surface. If this is central, the stress is very nearly equally distributed and shows few peaks; but when it is eccentrically placed within the available contact surface, the distribution is unequal and there are high stress peaks in the periphery. Pauwels (1965, 1980) was able to demonstrate this, impressively, in terms of the stress distribution in cases of coxa valga and vara. This variable distribution can be seen in normal anterior–posterior (AP) radiographs because of the different forms of the subchondral layer in the acetabulum.

Unequal stress distribution as a result of the eccentric position of a resultant force is not, however, necessarily pathological, but may also be a component of normal joint mechanics. This can be seen in the mineralization pattern of the tibial plateau in healthy subjects, where there is always a more highly mineralized maximum in the medial compartment; a fact which demonstrates that the pressure value there is, in general, greater than in the lateral compartment. The somewhat higher medial density is due to the slight medial displacement of the resultant force (Walker and Hajek 1972, Maquet 1976), so that the medial condyle is compelled to carry a greater load, probably because of the rather stronger ligamentous connections on the medial side. During flexion, the spatial extension of the attachments of the two collateral ligaments brings about a kind of twisted overlapping movement of their attachment zones which is somewhat reminiscent of stage "flats". This mechanism, which is caused by the integration of the hinter part of the medial meniscus with the dorsal component of the medial collateral ligament, is more marked and, therefore, leads to greater stability of the medial compartment. This also explains why the axis of rotation comes to lie rather closer to the medial compartment, which itself causes a significantly greater displacement of the lateral meniscus.

A similar mechanism is also present in the ankle joint. When interpreting the pattern found in this joint, it is, in our opinion, also necessary to be aware that the ankle joint does not only receive axial loading, but that with each step there is a turning or tilting of the foot which, as in a gear system, must lead to a characteristic sequential change in the contact surfaces. The stability of the ankle joint is principally due to the medial collateral ligaments. Since the lever arms of the medial ligaments are less than those of the lateral ligaments, and the constant stress on the deltoid ligament is responsible for the maintenance of stability and, therefore, for the medial position of the penetration point of the joint reaction force, it must follow that the higher pressure peaks occur here. With this fact, the higher density values in the region can be easily explained. In this joint, it must also be assumed that, because of the rather more marked ligamentous stabilization on the medial side, the resultant is slightly displaced in this direction.

Whereas a medial displacement of the resultant forces in the knee and ankle joints is always to be found in healthy subjects, results obtained from normal wrist joints reveal that the proportional transmission of force between the radial and lunate compartments differs in degree between individuals. According to An (1993), about 85% of the force transmitted by a healthy wrist joint in the neutral position is supposed to pass through the radius and the remaining 15% through the ulna compartment (Palmer and Werner 1984; Trumble et al. 1987). The scaphoid fovea of the lower articular surface of the radius, therefore, carries about 46–50% of the total force, the lunate fovea about 32–35%. The remainder is transmitted through the ulnocarpal compartment of the joint (Horii et al. 1990; Hara et al. 1992). Appropriately enough, our mineralization patterns show, in each case, a density maximum in the radial facets for scaphoid and lunate and just in that position where the highest stress has been measured during biomechanical experiments. Also in X-ray densitometric specimens, these maxima were found by Koebke et al. 1989, and Lehmann 1990.

A comparison of the peak values of the maxima in both compartments shows that, in the majority of cases ($n = 8$), the lunate fovea is frequently more highly mineralized and that it is here that the greater stress is active. This supports the belief that the lunate is to be regarded as the central building stone of the wrist joint and that, in normal cases, it transmits the greater part of the axial force to the forearm (Sennwald 1987). In four cases, the degree of mineralization was higher in the scaphoid fovea, in five, it was the same in both compartments. From this investigation, it can be concluded that not only one type of stress should be regarded as physiological. It is rather the case that the scaphoid and lunate types of mineralization as well as the equally balanced type should be regarded as normal variants.

7.5
Changes in the Mineralization Patterns With Age

Our results have also made it quite clear that an unambiguous age-dependent difference exists and, therefore, results and findings obtained from older people cannot be generalized and transferred to younger subjects. Differences observed when the density charts of the shoulder joints, hip joints and the talar component of the ankle joint from young and old people were compared have revealed that the mechanical conditions in these joints change with age. An X-ray densitometric study by Oberländer (1973) on sections of six specimens of the hip joint taken from the dissecting room, although constituting a small sample, have confirmed our findings in the living.

Changes in the geometry of the hip joint and also the humero-ulnar joint have been described by Bullough et al. (1973). The patterns of mineralization found by us in the humero-ulnar joint have allowed us to recognize only a limited age dependence, which is possibly because the elbow joints that we examined came mostly from the dissecting room, where the sample naturally consists mostly of older people.

The next step inevitably drives one to question what the cause for the age-dependent changes in the geometrical form of the joint may be. One can imagine many possibilities. For instance, the pattern of movement may alter and this would lead to a different stress distribution or there might be structural changes in the joint components. In our opinion, an altered pattern could certainly lead to quantitative changes in the mechanics of a joint, but for the qualitative changes which are here under discussion, alterations in the structural and mechanical characteristics of the tissue are a much more likely cause. Changes with age have, meanwhile, been described not only for the articular cartilages and bones, but also for the neuromuscular system, which plays an important role in joint protection. It is also conceivable that the different congruity relationships seen in different age groups may depend essentially on the altered viscoelastic properties of bone and cartilage. It is striking that such age-dependent changes have so far only been observed in joints in which a physiological incongruity has been postulated as the basis for adequate pressure transmission by the tissues (hip joint, shoulder joint, humero-ulnar joint and ankle joint). In the humeroradial, femorotibial and femoropatellar joints, no age-dependent differences in the mineralization pattern have been observed, provided one excludes changes such as those due to a malalignment of the knee joint.

8 Factors Influencing the Development of Pathological Patterns of Mineralization

The influence of the various mechanical factors on joint mechanics is even more clearly recognizable from the pattern of mineralization under pathological conditions than from the pattern seen under normal conditions. Here, one can observe "nature's experiments", so to speak, where the factors start to deviate from those present in good health, although the mutual interdependence is also inevitably influenced by other factors.

8.1
Abnormal Geometrical Relationships

Abnormal geometrical relationships are found in dysplastic hip joints. By this term one understands the complete or partial failure in development of the hip joint. The socket is flattened and its roof runs steeply instead of arching over the femoral head like a bow. The angle of the neck of the femur is also too steep and the degree of anterior torsion abnormal. The dysplastic roof of the socket does not cover the entire head of the femur and the unloaded part of the femoral head epiphysis grows more than the loaded part. These events are responsible for the typical shape of the head epiphysis (Putti 1937), which persists even after the cessation of growth and, like the dysplastic socket, is to be seen as a prearthrotic deformity. Owing to increasing dislocation of the femoral head and the steepness of the socket, the size of the bearing surface becomes progressively smaller. If the resultant force lies very near to the convexity of the acetabulum or if, because of deformity of the head, the transmission of force is contracted to a point, enormous pressure peaks arise in this region.

This transmission of force at a single point could be recognized without exception in all the cases of hip dysplasia examined by means of the pattern of mineralization in the lateral part of the roof of the socket, close to the region of greatest curvature. The high-density values are well above average and account for the enormous stress peaks already mentioned which, in course of time, certainly exceed the biological tolerance of the material and accelerate the development of arthrosis.

It is not surprising that, even with a low-grade scoliosis, the vertebral joints show different patterns on the right and left sides. These reflect the unequal stress resulting from the lateral curvature of the column in the frontal plane, which is accompanied by rotation about its long axis. The asymmetrical relationships inevitably produce an unequal distribution of stress within the motion segments.

8.2
Changes in the Magnitude of the Joint Reaction Force

As in the case of the physiological mineralization pattern, long-term changes in the size of the resultant force reveal themselves by their influence on the total pattern of mineralization observed in the patients examined.

After a short time, immobilized joint surfaces reveal a significant reduction in the total density, even while the position of the maxima remains unchanged. Further confirmation of the reduction in bone density, as an expression of long-term adaptation, was found in the glenoid cavities of two patients, in whom considerable delay in the reduction of shoulder dislocations led to complete removal of loading from the humeral head and its socket. The total mineralization was significantly reduced here, although the bicentric distribution pattern was not altered.

Significant increase in the force acting on a joint occurs in certain sports. In gymnastics at the rings, for instance, some phases of this activity produce forces in the shoulder joint of as much as five times the weight of the body (Bodem at al. 1984). The powerful forces inevitably lead to greatly increased pressure on the joint, which expresses itself in the subchondral pattern by an excessive increase in the total mineralization. It is theoretically conceivable that the active joint pressures reached may exceed the level tolerated by the articular cartilage. The patterns observed do not, however, allow one to decide exactly where the borderline might lie between increased physiological mineralization in response to a greatly increased compressive stress and a pathological rise in density which might be associated with subsequent arthrosis.

8.3
Changes in Size and Position of the Contact Surfaces

The great importance of the menisci in providing an adequate distribution of stress for the tissue in the articular surface of the knee joint has been emphasized by many authors. The meniscus determines decisively the stress distribution in the joint surface both directly by its influence on the location and size of the contact surfaces (Ahmed and Burke 1983; Baratz et al. 1986, Fairbank 1948; Fukubayashi and Kurosawa 1980; Maquet et al. 1975; Kummer 1987) and indirectly by its influence on the extent and position of the joint reaction force by extending the area available for the penetration of the resultant.

Owing to a reduction in the contact surfaces to about half their normal size and an increase in the value of the resultant force following meniscectomy, the local stress in the joint surface rises. The resulting forces (pressure below the contact surfaces, tension at the periphery) bring about an unphysiological deformation of the subchondral plate and the underlying trabeculae. Furthermore, as pressure measurements made on cadaveric specimens by Walker and Erkmann (1975) and Ahmed and Burke (1983) have shown, there is a small displacement of the stress peaks toward the periphery after meniscectomy. Since, within physiological limits, bone reacts to the degree of stress to which it is subjected by degradation or deposition (Kummer 1985; Pauwels 1965, 1980), the subchondral mineralization pattern, which is normally less stressed because of the overlying meniscus, must undergo an increase in density and thickness, particularly at the edges.

In fact, there is a significant displacement, medially and dorsally, of the density maxima in the medial joint compartment from the central position, as our investigations following excision of the medial meniscus in sheep have demonstrated. This agrees with the findings of Odgaard et al. (1989) and Noble and Alexander (1985), who also observed displacement of the density after meniscectomy in a retrospective study of human subjects. In the lateral compartment of the joint, on the other hand, there is indeed no such significant displacement of the maxima, although the position of these density maxima, which is not so constant as on the normal control side, can be regarded as a result of the rotational instability and, therefore, an altered stress on the joint surface. This finding indirectly confirms the results obtained by Wang and Walker (1974), as a consequence of which, the menisci can be regarded as not unimportant secondary stabilizers of rotation.

One disputes that the cruciate ligaments play a large part in maintaining the stability of the knee joint. Because of their passive control of movement here, they not only have a decisive influence on the position of the contact surfaces, but they also exercise an important function in activating the periarticular musculature and, thus, indirectly affect the position of the resultant force. The basis of this is the presence of mechanoreceptors in the anterior cruciate ligament, which are predominantly situated near to its femoral attachment (Haus and Halata 1990), and from which stretch-dependent signals are transmitted to the spinal cord (Krauspe et al. 1993). After these have been processed and, where necessary, subjected to central modulation, the corresponding muscles are activated accordingly.

It is relevant here to observe that, after the loss of proprioception brought about by rupture of the anterior cruciate, the reflex arc between the ligament and the flexor muscles (LCA reflex) is interrupted (Beard et al. 1993; Corrigan et al. 1992; Grüber et al. 1986; Solomonow et al. 1987). The poor coordination of the periarticular musculature, together with failure of the mechanical stabilization of the joint, may be regarded as the cause of the displacement of the penetration point of the resultant force and this leads to a pathological distribution of stress in the joint which must, in turn, produce a redistribution of the subchondral mineralization. In the same way, the animal model showed, particularly in the medial compartment, a displacement of the mineralization maxima in the tibial plateau towards the mediodorsal border after the anterior cruciate ligament had been divided.

8.4
Changes in the Penetration Point of the Joint Reaction Force

It must be accepted that it is not only an increase or decrease in activity that leads to a change in the subchondral density pattern, but also a displacement of the resultant force, indicating unphysiological loading on the joint. That such changes in adaptation can follow the alteration of mechanical conditions within a joint had already been demonstrated in AP radiographs by Pauwels (1973). Following a correction osteotomy for coxa valga or coxa vara, the form and extent of the subchondral bone at the hip joint changes with time.

Möllers et al. (1986) carried out radiodensitometric investigations on the distribution of the subchondral bone density at the lower end of the radius in normal subjects and in cases of defectively reduced Colles' fractures. The distribution of the material in the normal radius followed the typical pattern with two central maxima, which differed widely from that of the pathological cases. The abnormal pattern must, particularly because of the dorsal displacement of the density maxima, be interpreted as the expression of an altered distribution of stress over the distal surface of the radius, with increased loading of the dorsal section. This can account for the frequently encountered pain and risk of subsequent arthrosis. This dorsal displacement of the maxima was similarly encountered in the patients examined by us.

In a combined study with Frahm from the Radiology Department at the University Hospital of Freiburg, it has been possible to confirm Möllers' (1986) findings on patients by OAM. The subchondral pattern of mineralization in patients with defectively repositioned healed fractures of the distal end of the radius showed typical displacements of the density maxima, associated with variably extensive defective repositioning at different degrees of pronation or supination and dependent on the angle of deviation.

The relationship between stress distribution changes because of the altered position of the point of penetration of the resultant force and the mineralization pattern is strikingly demonstrated by the results of a malalignment of the knee joint. In normal cases, the density maxima on both the lateral and medial sides indicate a fairly equal distribution of stress between the two tibial condyles. A somewhat higher density on the medial side comes about because the point at which the resultant force is transmitted in this joint is not exactly in the middle, but somewhat displaced medially. For this reason, the medial condyle transmits a somewhat heavier load (Walker and Hajek 1972; Maquet et al. 1975).

In cases of genu valgum, there is a marked increase in the mineralization of the lateral part of the tibial plateau, while the medial side presents a significantly reduced bone density. It is especially the lateral displacement of the axis described by Maquet (1976) that leads to increased stress on the lateral part of the joint, together with a simultaneous unloading of the medial joint surface. With genu varum, the opposite effect arises, owing to displacement of the knee-joint axis medially. The greatly increased stress in the peripheral region of the medial tibial condyle leads to the development of a medial peripheral density maximum, while the reduced lateral stress produces a corresponding reduction of the mineralization in the lateral part of the joint. The additional inclusion of the femoral condyles reflects this situation, which signifies that all the components of the femorotibial joint react in a like manner to changes in the mechanical relationships.

The mineralization patterns which we have observed in detail reflect these relationships and confirm the accuracy of the stress diagram put forward by Maquet (1976).

A similar pathological mechanism must be assumed for certain diseases of the shoulder joint. The type of picture with a monocentric dorsal maximum is only found in patients with various forms of instability or in those with recurrent dislocation (Müller-Gerbl et al. 1993a). The observed displacement of the zone of greatest density to the dorsal periphery of the joint surface, in patients with the above conditions, seems to suggest that the cause lies in the long-term eccentric positioning of the resultant force and, with this, the unequal distribution of stress throughout the joint surface. In this way, the peripheral peak values are produced and these eventually damage the cartilage and bone and, thus, encourage the development of arthrosis. Pain syndromes unaccompanied by any corresponding morphological changes may be explained in this way.

106

The cause for the long-term eccentric displacement of the resultant force may be sought in the interaction of the muscles of the so-called rotator cuff, which exerts considerable influence on the positioning of the humeral head in its socket. Whereas the mineralization patterns observed with the various types of instability must lead one to assume a pathological displacement in the ventrodorsal direction, in patients with rupture of the rotator cuff, one often observes a pattern indicating displacement of the zone of greatest density in a cranial direction and this can account for the high position of the humeral head, resulting from there being insufficient muscular force to keep it in the center of the socket.

8.5
The Temporal Course of the Changes

The large number of results so far obtained strikingly illustrate that the subchondral mineralization patterns are a reflection of the individual stresses acting on a joint and that this is well supported by current models of joint mechanics. However, how long it takes for the pattern to become remodelled in adaptation to the changed mechanical conditions remains an open question.

In order to answer this question, we have started a prospective study of the density patterns in patients with genu varum to compare the patterns before and 1 year after a correction osteotomy. The basic hypothesis is that a displacement of the resultant force towards the center, brought about by correction of the axis, must eventually produce an equal distribution of stress in the two compartments and, thus, bring about a corresponding displacement of the density maxima.

One year after the correction osteotomy there was a significant alteration in the density patterns in comparison with how they were before the operation. Apart from a more or less marked reduction in the general mineralization, which is almost certainly due to postoperative immobilization, a distinction between the results can be made in terms of the parameters "displacement" and "alteration in density". In the patients of group 1, both findings (displacement of the medial peripheral maximum together with increased mineralization laterally) indicate that the resultant force was successfully shifted in a lateral direction.

The significantly higher initial mineralization in group 2, together with the more marked postoperative reduction of the high density level, suggests that the time taken for the displacements to occur is fundamentally dependent on the degree of preoperative mineralization. It is to be assumed that examination of these patients after more than 1 year following the operation must disclose similar displacements to those found in group 1.

The postoperative findings in groups 3 and 4, on the other hand, must be regarded, from our point of view, as revealing a deterioration in the mechanical conditions since they, unlike the former, indicate an increase in the stress distribution which is unfavorable for genu varum. The differences in our results accord with clinical experience that not every osteotomy produces the desired result.

With this comparative study it has been possible, for the first time, to establish that changes in the subchondral density are an adaptation to an altered mechanical situation in human subjects, and we have good evidence for thinking that this is based on a change in the local stress. However, in order to prove that the theoretical concept underlying the correction osteotomy (to "valgarize" genu varum and, by shifting the stress laterally, produce an equal distribution within the tibial plateau) is relevant, the extent of the change in the mechanics must be tested by an exact analysis of the individual patient.

During a further investigation into the changes in mineralization after osteotomy, which took into account the extent in depth, a highly significant (p <0.01) regression of the density values in the first ten layers emerged. In the postoperative knee, the maximal pixel count in all the HU stages became displaced into deeper layers than in the preoperative situation (an average of 4 sections, p <0.01) and, in so doing, the position of the density maxima and their extent in depth approached that of a normal knee. In other words, in cases of genu varum treated by correction osteotomy, there was a vertical shift of the higher density values into the deeper regions as well as a horizontal displacement of the density maxima within the joint surface, both of which indicate an approach towards conditions of normality. Putting this the other way round, it means that, in cases of genu varum, bone remodelling takes place close to the cartilage. In terms of the theory that the density of the tissues is modified by the local stress, we present the hypothesis that the beginning of arthrotic remodelling takes place in the layers nearest the cartilage and is determined by the more superficial stress peaks.

9 Possible Clinical Applications of CT OAM

9.1
Basic Clinical Research

Apart from biochemical, endocrine, metabolic or genetic causes of arthrosis, it is most especially the degenerative joint changes of mechanical origin which, because of ever-increasing age and the modified demands of professional and free-time activities, are responsible for the continually rising number of patients and, of course, the consequent economic effects. It can be assumed that arthrosis is due to a faulty relationship between the loading in a joint, on the one hand, and the ability of its components to support that load, on the other.

A hint as to the etiology and pathogenesis of a variety of joint diseases has been provided by the pattern of mineralization in the femoropatellar joint. The density patterns found in normal subjects can be correlated with details of the contact surfaces reported in the literature (Hehne 1983). They also agree with the radiodensitometric findings reported by Tillmann (1980). Here, there is also a relationship between the distribution of the subchondral density and that of the pressure on the joint surface, and the pressure distribution itself is, according to present views, dependent on the antagonistic interaction between the vastus medialis and other components of the quadriceps (Müller 1982; Maquet 1984).

Particularly interesting findings have been obtained from patients with damaged cartilage, in whom the position of the density maxima in the proximal part of the lateral patellar facet is the same as in healthy subjects. With damage to the lateral cartilage, the total mineralization is significantly raised as the damage becomes more serious, but with damage to the medial cartilage it is strongly reduced and, this, taking into account the mechanics of the knee joint, seems to imply different etiologies for the two types of damage (Eckstein et al. 1993c, d).

Analogous to the considerations concerning stress distribution in the acetabular socket, our opinion is that damage to the medial cartilage is to be interpreted as the expression of a long-term reduction of the demands upon a particular area of cartilage, whereas the damage to the lateral cartilage is the result of a continued high loading in this region. It is suggested that lateral cartilage damage is the result of an unequal distribution of the pressure with peak stress values which exceed those that can be tolerated by the cartilage.

Degenerative changes in cartilage can, therefore, follow from two different mechanical situations: either from a continued heavy loading of the tissue or from insufficient demand being made upon it. According to Mohr (1984), the initial changes of chondromalacia patellae, which involves the medial joint facet, lead to complete disappearance of the cartilage, with both its light and electron microscopic changes (Paul 1975) showing a picture very similar to the arthrotic cartilage of other joints. This means that, independent of the mechanical causes, the articular cartilage can only react to these disturbances in a particular predetermined manner. Whether, because of their different etiologies, both kinds of cartilaginous damage show any difference in behavior as they progress – even though they present the same morpho-pathological and biochemical picture (Mohr 1984) - is an important question that still remains to be answered. Investigations carried out by Eckstein (1993b) have suggested that the progress of cartilaginous damage above highly mineralized areas takes place more rapidly than it does above the less heavily mineralized zones.

Age-dependent, degenerative changes in hip-joint cartilage begin in unloaded rather than in loaded parts of the joint (Harrison et al. 1953; Byers et al. 1970; Bullough et al. 1973). They involve a more or less triangular zone in the region of the roof of the socket, the basis of which lies along the outer edge (Bullough et al. 1973). The alterations range from a reduced proteoglycan content to wearing away of the cartilage (fibrillation of the cartilage – Bullough et al. 1968). A reduction in the amount of proteoglycan can result from many causes – a fact which has, so far, remained largely unobserved – including insufficient use or permanent reduction in the loading. In the same way that immobilized bone or resting muscle atrophies, insufficient demands upon cartilage can produce a reduction in proteoglycan synthesis by the chondrocytes. It has been observed in the elbow joint by Goodfellow and Bullough (1967) and in the knee joint by Goodfellow et al. (1971). that there is a positive correlation between regions in which degeneration is often found and those which, as a matter of course, scarcely support any load. These degeneratively altered, chondromalacial regions of cartilage do not impair joint function so long as the incongruity is maintained. In older people, however, this articular incongruity, which guarantees an acceptable physiological environment, is lost, resulting in an altered distribution of pressure in the joint, so that gradually a quantitative increase in the pressure begins to act on previously unloaded zones. These regions of the joint, which have become chondromalacial as a result of a long-term reduction in use, now have to carry a severe stress; one to which they have become poorly adapted. In this way, a process of cartilage deterioration which cannot be halted is begun and is, in turn, accelerated by the reduced nourishment of the tissues. It means that the permanent underactivity of these regions of cartilage, together with a reduction in incongruity, is an important factor in the development of mechanically induced arthrosis.

A defined alteration in the mechanical situation of a joint, leading to changes in the stress distribution and, later, to arthrosis, can be demonstrated in the animal model after the removal of a meniscus, the essential significance of which for an adequate stress distribution throughout the tissues has been very thoroughly investigated. A year after the operation, all the animals examined show a displacement of the density maxima toward the periphery, with extensive cartilaginous lesions. Furthermore, increase in the thickness and density of the subchondral mineralization zone and the subarticular trabeculae, as a consequence of the altered stress or as a result of bone remodelling, can be demonstrated (Anetzberger et al. 1995). Only the mineralization patterns can be obtained in vivo – and without interference – and both the questions asked and the sample itself can be selected so as to assess the role of long-term stress acting on the articular surface as the cause of a mechanically induced arthrosis.

Meniscectomy in sheep also provides a suitable model of arthrosis, since the degenerative changes appear very early.

Interesting possibilities are suggested by investigations into the mineralization patterns in the wrist joint in patients with Kienböck's disease, the etiology of which has given rise to a great deal of controversial discussion. Most authors consider the cause to be an alteration in the transmission of pressure through the joint. The resulting severe stress on the lunate – often accompanied by an ulna minus wrist – is considered to lead, in the course of time, to an avascular necrosis. Instead of the expected high mineralization peaks in the lunate fossa and the appearance of only a single type of loading on the lunate, all three types of loading appear. In advanced stages of the disease, no maximum can be demonstrated in the lunate, probably because of the collapse of this bone and the consequent failure of pressure transmission.

It is striking that these patients show, compared with the normal subject, significantly reduced total mineralization, not only in the affected joint, but also on the contralateral side. In addition to the biometric results obtained by Tsuge and Nakamura (1993), who also identified the specific configuration of the lunate in the unaffected hand, the mineralization patterns found by us have suggested that there is a genetic predisposition to Kienböck's disease which is triggered by some factor and so becomes clinically manifested. Such a hypothesis is supported by the clinical observation that some patients eventually develop bilateral avascular necrosis of the lunate.

9.2
Diagnosis

Since the pattern of distribution of the subchondral mineralization allows quantitative and qualitative conclusions to be drawn about the principal stress, it can be used as a noninvasive diagnostic procedure in selected cases where long-term unphysiological stress is suspected.

It is very interesting that the shoulder-joint sockets of top gymnasts revealed, in two-thirds of those examined, a pattern in which the zone of greatest density was found in the dorsal periphery of the joint; a pattern that is never found in ordinary subjects and which allows one to conclude that the position of the resultant force must be a predominantly eccentric one. Since these athletes complained of pain in the region of the acromioclavicular joint, its articular surfaces were also included in the investigation (Müller-Gerbl et al. 1990a). Whereas a central position for the joint reaction force in the shoulder joint leads to a central penetration point of joint reaction force in the AC joint (centrally located density maximum), in gymnasts with a dorsal maximum in the glenoid cavity, a dorsal maximum was also found in the AC joint. The findings in both joints indicate an unphysiological stress which can lead to overloading of certain parts of the joint and also possibly result in pain if the level of tolerance of the bone and cartilage is permanently exceeded, even though the actual cause of this pain is still unclear. This unphysiological stress, which arises in certain athletes, may possibly be due to insufficient muscle training or to an inadequate technique. The method could, therefore, also be used in the field of sport medicine or for assessing professional injuries or a reduction in the capacity for work.

Unphysiological stress can also be related to impaired coordination of the periarticular musculature, whether this is due to paralysis of an isolated muscle, muscular imbalance or a disturbance of the neuromuscular control, and can lead to an eccentric distribution of stress in the affected joint. Although the relationship is obvious enough in theory, it requires proof and this can be obtained by CT OAM, provided that possible harm caused by the necessary irradiation is taken into account.

9.3
Following the Progress of Treatment

CT OAM is particularly well suited to the follow-up of various therapeutic procedures which are directed toward bringing about changes in the mechanics of a joint. The quickest changes are to be expected after a correction osteotomy to shift the axis.

Examinations made after treatment of genu varum by tibial osteotomies show, after a successful operation, density patterns which correspond to the theoretical expectations. Following lateral displacement of the axis, a redistribution of the load takes place between the medial and lateral parts of the tibial plateau, revealed by a reduction and a central displacement of the high peripheral density maxima on the medial side and accompanied by an increase in the density on the lateral side. The improved stress distribution achieved by this operation should also bring about an improvement in the arthrotic condition. Corresponding to clinical experience, which teaches us that the results of the operation are not always satisfactory, postoperative density patterns are also found that indicate a deterioration in the condition.

The result of a reconstruction of the anterior cruciate ligament is in general assessed by clinical and biomechanical examination. As the postoperative interval grows longer, there appears in many cases instability, eventually followed by arthrosis. Apparently, it is not always possible to restore the physiological function of the cruciates surgically. On theoretical grounds alone, this must finally lead to a pathological distribution of stress within the joint. This is itself likely to produce the laying down of an altered distribution of the subchondral density in accordance with Wolff's law. We were able, by means of CT OAM, to demonstrate in sheep that a pathological alteration in the stress distribution can be demonstrated by characteristic changes in the mineralization of the subchondral bone within a year of replacing the anterior cruciate. CT OAM offers the possibility of assessing the influence of various factors (the operative technique, histological and biomechanical characteristics of the transplant, pretension of the transplant, position of the drill holes, postoperative treatment etc.) on the stress distribution and, thus, of checking the results of the operation and eventually developing a better technique.

The high incidence of arthrosis following meniscectomy has produced a large variety of ways of carrying out replacement of the meniscus, the results of which are assessed in terms of morphology and mechanics. Whereas a significant displacement of the medial density maximum toward the mediodorsal edge is found after removal of the medial meniscus, replacement of the meniscus by an autologous transplant from the patellar tendon covered with fascia lata is followed by only a dorsal shift of the maximum. The results of replacing the meniscus have shown that there is a definite improvement in the mechanical situation, even if this cannot always be achieved. Evaluation of the subchondral mineralization by CT OAM allows an objective analysis of the whole course of mechanically effective operations on the joint to be made, for which purpose it is much better than an ordinary AP radiograph that, as a summation picture, can only offer a rough estimation of the subchondral density. With the former method, an exact objective localization of the density distribution is possible. In other words, a method has been made available which can be used in the living subject, and which makes it possible to draw conclusions about the current situation within an individual joint and, so, allows the postoperative course of the treatment to be quantified.

10 Summary

Pauwels (1965) and subsequent workers in the same field have shown that the distribution of the subchondral density within a joint surface can serve as a parametric measurement which reflects the main stress acting on a joint. Our own investigations on anatomical specimens have demonstrated that this subchondral mineralization does indeed show regular distribution patterns from which conclusions about the mechanical situation within an individual joint may be drawn. Since radiographical densitometry and histological methods are only available for determining the adaptive reaction of the bone to the particular mechanical situation in a joint after death, the information obtained applies only to an end situation and tells us nothing about the development of the changes with time. Furthermore, investigations carried out on human specimens by radiographical densitometry mostly apply to samples of a particular age, since such specimens can be acquired only from departments of pathology, forensic medicine or anatomy. The functional reactions of the bone tissue to repeated long-term changes in the loading – lengthy immobilization and subsequent remobilization, for instance, or heavy loading over a considerable period of time – cannot be followed by any ordinary method in experimental animals, since the death of the animal is a prerequisite for the precise quantitative examination of the bone tissue. This applies also to attempts to follow the process by means of animal experiments.

CT OAM has been developed as a method which, based on CT, can provide a surface representation of the 3-D density distribution in the joints of living subjects.

Comparative studies were carried out to establish and confirm the validity of the procedure. These have shown (1) that the results obtained from anatomical specimens are identical with those obtained in the living; (2) that secondary CT sections are suitable for evaluation and that the spectrum of joint surfaces examined can be extended to include the whole joint (if this were not so, effects caused by the apparatus – particularly the partial-volume effect – would render the procedure impossible); and finally (3) that the distribution of the Hounsfield density within the subchondral bone represents the distribution of the mineralization.

The mineralization patterns found by us in different joints of normal subjects have shown that these patterns can be brought into line with current models of joint mechanics. The radiocarpal joint, for instance, has revealed the various types of loading occurring within physiological limits. Information has also been obtained about the age-related changes taking place in the hip, wrist and ankle joints.

The increase of the total mineralization in gymnasts can be related to the qualitative and quantitative adaptation to an increased peak loading, and reduced mineralization to a lengthy reduction in use during, for instance, postoperative immobilization.

In groups of patients with various diseases of mechanical origin (shoulder instability, malalignment of the main axis, defective repositioning of healed fractures, rupture of the rotator cuff, meniscectomy or rupture of the anterior cruciate ligament), a pattern of mineralization is found which is different from the normal picture. These findings reflect the abnormal mechanical situation.

The mineralization pattern of the femoropatellar joint has revealed the differing etiologies of medial and lateral cartilage damage and the examination of patients with lunatomalacia has made it possible to recognize a genetic disposition.

The postoperative comparison of the mineralization patterns of patients with genu varum who have undergone a correction osteotomy and the results of animal experiments on various procedures for reconstructing the anterior cruciate ligament or a primary replacement of the meniscus, have produced results which make it possible to judge the success or failure of the operation. Since subchondral mineralization, as a correlate of the main stress in a joint, can be assessed in the living by means of CT OAM, this procedure offers a method with a wide spectrum of applications, provided the possible sources of error are taken into account. Besides the specific uses that CT OAM has made available in the clinical field, the method offers a useful contribution to basic research, particularly since, with the new method of 3-D reconstruction, the time required is much less.

Since it is a noninvasive method of examination which is not unduly trying for the patient, CT OAM can be used as a diagnostic procedure to provide a morphological answer to the effect of long-term stress on the mechanical conditions of an individual joint. In particular, estimation of the position of the joint reaction force or the "age of the joint" is possible. Since alterations in the mineralization can be attributed to the local cells (osteocytes, osteoblasts, chondrocytes, chondroblasts, fibrocytes, fibroblasts), CT OAM makes it possible to discover what mechanical influences can lead to metabolic changes in the subchondral bone.

Because CT OAM has provided a means for interpreting the exact mechanical conditions within small regions of a joint, it is a method which, in our opinion, can be used for examining those whose joints are particularly liable to stress, either at work or in the field of sport, with a view to recognizing damage early. It can also be used for the objective evaluation of prearthrotic changes which, because of the large number of CT machines available, need not be confined to the occasional patient.

In view of the increasing number of patients who are suffering from degenerative joint changes, CT OAM should be able to play an important role in research on the disease and the diagnosis of arthrosis and is particularly useful for distinguishing between the early stages and different forms of prearthrosis. Apart from this, the method makes it possible to follow the course of all operations on joints which are of mechanical significance; correction osteotomy for coxa valga or vara, for instance, or operations for reducing the load on the femoropatellar joint. This applies also to retrospective studies, since practically all CT datasets with a small slice thickness can be taken from the archives and used for this purpose.

References

Adams MA, Hutton WC (1983) The mechanical function of the lumbar apophyseal joints. Spine 8:327–330

Adams J, Chen S, Adams P, Isherwood I (1982) Measurement of trabecular bone mineral by dual energy computed tomography. J Comput Assist Tomogr 6:601–607

Afoke NYW, Bryers PD, Hutton WC (1980) The incongruous hip-joint: a casting study. J Bone Joint Surg Br 69:511–514

Ahmed AM, Burke DL (1983) In vitro measurement of static pressure distribution in synovial joints–Part I: tibial surface of the knee. J Biomech Eng 105:216–225

Alexander H, Ricci JL, Grande DA, Blumenthal NC (1984) Mechanisms of stress-induced remodeling. Interne Veröffentlichung: H Alexander, Department of Bioengineering, Hospital for Joint Diseases Orthopaedic Institute, New York, NY 10003

Alvarez RE, Macovski A (1976) Energy-selective reconstruction in X-ray computerized tomography. Phys Med Biol 21:733–744

Amir G, Pirie CJ, Rashad S, Revell PA (1992) Remodelling of subchondral bone in osteoarthritis: a histomorphometric study. J Clin Pathol 45(11):990–992

Amtmann E (1971) Mechanical stress, functional adaption and the variation-structure of the femur diaphysis. Ergebn Anat Entwickl-Gesch 44:1–89

Amtmann E, Schmitt HP (1968) Über die Verteilung der Corticalisdichte im menschlichen Femurschaft und ihre Bedeutung für die Bestimmung der Knochenfestigkeit. Z Anat Entwickl-Gesch 127:25–41

An KN (1993) The effect of force transmission on the carpus after procedures used to treat Kienböck's disease. Hand Clin 9:445–454

Andriacchi TP, Stanwyck TS, Galante JO (1986) Knee biomechanics and total knee replacement. J Arthroplasty 1:211–219

Anetzberger H, Putz R (1996) The scapula: principles of construction and stress. Acta Anat 156:70–80

Anetzberger H, Müller-Gerbl M, Scherer MA, Metak G, Blumel G, Putz R (1994) Veränderung der subchondralen Mineralisierung nach Rekonstruktion des vorderen Kreuzbands beim Schaf. Unfallchirurg 97:655–660

Anetzberger H, Metak G, Scherer MA, Putz R, Müller-Gerbl M (1995) Anpassung der subchondralen Knochenplatte nach Meniskektomie als Folge einer Änderung der Spannungsverteilung. Osteologie 4:224–232

Ascenzi A, Bonucci E (1964) The ultimate tensile strength of single osteons. Acta Anat 58:160–183

Athesian GA, Lai WM, Zhu WB, Mow VC (1994) An asymptotic solution for the contact of two biphasic cartilage layers. J Biomech 27:1437–1460

Baratz ME, Fu FH, Mengato R (1986) Meniscal tears: the effect of meniscectomy and of repair on intraarticular contact areas and stress in the human knee. A preliminary report. Am J Sports Med 14:270–275

Beard DJ, Kyberd PJ, Fergusson CM, Dodd CAF (1993) Proprioception after rupture of the anterior cruciate ligament. J.Bone Joint Surg Br 75-B:311–315

Behrens JC, Walker PS, Shoji H (1974) Variations in strength and structure of cancellous bone at the knee. J Biomech 7:201–207

Bennett GA, Waine H, Bauer W (1942) Changes in the knee joint at various ages. In: The Commonwealth Fund, New York, pp1–51

Benske J, Schunke M, Tillmann B (1988) Subchondral bone formation in arthrosis. Polychrome labeling studies in mice. Acta Orthop Scand 59:536–541

Bentzen SM, Hvid I, Jorgensen J (1987) Mechanical strength of tibial trabecular bone evaluated by X-ray. J Biomech 20:743–752

Berry JL, Thaeler-Oberdoerster DA, Greenwald AS (1986) Subchondral pathways to the superior surface of the human talus. Foot Ankle 7:2–9

Bindermann I, Shimshoni Z, Somjen D (1984) Biochemical pathways involved in the translation of physical stimulus into biological message. Calcif Tissue Int 36:582–585

Bjorkstrom S, Goldie IF (1982) Hardness of the subchondral bone of the patella in the normal state, in chondromalacia, and in osteoarthrosis. Acta Orthop Scand 53:451–462

Bland JH (1983) The reversibility of osteoarthritis: a review. Am J Med 74:16–26

Bodem F, Brussatis F, Menke W (1984) Zur theoretischen Biomechanik des Schultergelenkes: die Entstehung gewöhnlicher und außergewöhnlicher mechanischer Belastungen des glenohumeralen Gelenkknorpels. In: Refior HJ, Pilitz W, Jäger M, Hackenbroch (eds) Biomechanik der gesunden und kranken Schulter, Thieme, Stuttgart, pp82–87

Braus H (1920) Anatomie des Menschen, 1.Band Bewegungsapparat. Springer, Heidelberg

Brinell JA (1900) II Cong Int d Méthodes d'essay, Paris

Brown TD, Vrahas MS (1984) The apparent elastic modulus of the juxtarticular subchondral bone of the femoral head. J Orthop Res 2:32–38

Brown RA, Weiss JB (1988) Neovascularisation and its role in the osteoarthritic process. Ann Rheum Dis 47:881–885

Brown TD, Pedersen DR, Radin EL, Rose RM (1988) Global mechanical consequences of reduced cement/bone coupling. J Biomech 21:115–129

Bullough PG (1981) The geometry of diarthrodial joints, its physiologic maintenance, and the possible significance of age-related changes in the geometry-to-load distribution and the development of osteoarthritis. Clin Orthop 156:61–66

Bullough PG, Goodfellow JW, Greenwald AS, O'Connor J (1968) Incongruent surfaces in the human hip joint. Nature 217:1290

Bullough PG, Goodfellow, JW, O'Connor J (1973) The relationship between degenerative changes and load-bearing in the human hip. J Bone Joint Surg Br 55:746–758

Bullough PG, Yawitz PS, Tafra L, Boskey AL (1985) Topographical variations in the morphology and biochemistry of adult canine tibial plateau articular cartilage. J Orthop Res 3:1–16

Bunck S (1990) Krümmungs- und Kontaktflächenverhaltnisse der Articulatio humeroradialis. Anat Anz 171:45–53

Burger EH, Veldhuijzen JP (1993) Influence of mechanical factors on bone formation, resorption and growth in vitro. In: Hall BK (ed) Bone: a treatise, Vol 7, CRC Press, Boca Raton, pp37–65

Burr DB, Martin RB, Schaffler MB, Radin EL (1985) Bone remodeling in response to in vivo fatigue microdamage. J Biomech 18:189–200

Byers PD, Contepomi CA, Farkas TA (1970) A post mortem study of the hip joint. Ann Rheum Dis 29:15–31

Cameron HU, Fornasier VL (1979) Fine detail radiography of the femoral head in osteoarthritis. J Rheumatol 6:178–184

Canalis E, McCarthy T, Centrella M (1988) Growth factors and the regulation of bone remodeling. J Clin Invest 81:277–281

Cann CE (1988) Quantitative CT for determination of bone mineral density: a review. Radiology 166:509–522

Carter DR (1984) Mechanical loading histories and cortical bone remodeling. Calcif Tissue Int 36[Suppl]:19–24

Carter DR (1987) Mechanical loading history and skeletal biology. J Biomech 12:1095–1109

Carter DR, Hayes WC (1976) Bone compressive strength: the influence of density and strain. Science 194:1174–1176

Carter DR, Hayes WC (1977) The compressive behavior of bone as a two-phase porous structure. J Bone Joint Surg Am 59:954–962

Carter DR, Spengler DM (1978) Mechanical properties and composition of cortical bone. Clin Orthop 135:192–217

Carter DR, Schwab GH, Spengler DM (1980) Tensile fracture of cancellous bone. Acta Orthop Scand 51:733–741

Carter DR, Fyhrie DP, Whalen RT (1987) Trabecular bone density and loading history: regulation of connective tissue biology by mechanical energy. J Biomech 20:785–794

Carter DR, Orr TE, Fyhrie DP (1989) Relationship between loading history and femoral cancellous bone architecture. J Biomech 22:231–244

Chamay A, Tschantz P (1972) Mechanical influences in bone remodeling. Experimental research on Wolff's law. J Biomech 5:173–180

Choi K, Kuhn JL, Ciarelli MJ, Goldstein SA (1990) The elastic moduli of human subchondral, trabecular, and cortical bone tissue and the size-dependency of cortical bone modulus. J Biomech 23:1103–1113

Christensen P, Kjaer J, Melsen F, Nielsen HE, Sneppen O, Vang PS (1982) The subchondral bone of the proximal tibial epiphysis in osteoarthritis of the knee. Acta Orthop Scand 53:889–895

Clark JM, Huber JD (1990) The structure of the human subchondral plate. J Bone Joint Surg Br 72:866–873

Cochran G (1988) Orthopädische Biomechanik. Enke, Stuttgart

Corrigan JP, Cashman WF, Brady MP (1992) Proprioception in the cruciate deficient knee. J. Bone Joint Surg Br 74-B:247–250

Cowin SC (1986) Wolff's law of trabecular architecture at remodeling equilibrium. J Biomech Eng 108:83–88

Cowin SC, Hegedus DH (1976) Bone remodeling I: theory of adaptive elasticity. J Elasticity 6:313–326

Cowin SC, Weinbaum S, Zeng Y (1995) A case for bone canaliculi as the anatomical site of strain generated potentials. J Biomech 28:1281–1297

Currey JD (1970) The mechanical properties of bone. Clin Orthop 73:209–231

Dalstra M (1993) Biomechanical aspects of the pelvic bone and design criteria for acetabular prostheses. (thesis), University of Nijmegen, The Netherlands

Darracott J, Vernon-Roberts B (1971) The bony changes in "chondromalacia patellae". Rheumatol Phys Med 11:175–179

Davidovitch Z, Shanfeld JL (1975) Cyclic AMP levels in alveolar bone of orthodontically-treated cats. Arch Oral Biol 20:567–574

Davis MA (1988) Epidemiology of osteoarthritis. Clin Geriatr Med 4:241–255

Dedrick DK, Brandt K, Goulet RW (1990) Subchondral plate and trabecular bone in experimentally induced osteoarthritis. Arthritis Rheum 33[Suppl]:91

Dekel S, Weissman SL (1978) Joint changes after overuse and peak overloading of rabbit knees. Acta Orthop Scand 49:519–528

Deutsch AL, Mink JH, Waxman AD (1989) Occult fractures of the proximal femur: MR imaging. Radiology 170:113–116

Dewire P, Simkin PA (1996) Subchondral plate thickness reflects tensile stress in the primate. J Orthop Res 14:838–841

Dittmar O (1930) Beobachtungen an den Gelenkfortsätzen der Lendenwirbel bei sagittal- und lateralflexorischer Bewegung. Zur Mechanologie der Wirbelsäule. II. Mitteilung. Z Anat Entw Gesch 93:477–483

Donaldson C, Hulley S, Vogel J, Hattner R, Bayers J, McMillan D (1970) Effects of prolonged bed rest on bone mineral. Metabolism 19:1071–1084

Donohue JM, Buss D, Oegema TR Jr, Thompson RC Jr (1983) The effects of indirect blunt trauma on adult canine articular. J Bone Joint Surg Am 65:948–957

Ducheyne P, Heymans L, Martens M, Aernoudt E, de-Meester P, Mulier- P (1977) The mechanical behaviour of intracondylar cancellous bone of the. J Biomech 10:747–762

Dulce HJ (1970) Biochemie des Knochens. In: Diethelm L (ed) Handbuch der Medizinischen Radiologie IV/1, Springer, Berlin

Duncan H, Riddle JM, Jundt JW, Pitchford W (1985) Osteoarthritis and the subchondral plate. In: Verbruggen G, Veys EM (eds) Degenerative joints. Vol 2. Elsevier Science, Amsterdam, pp 181–197

Duncan H, Jundt J, Riddle JM, Pitchford W, Christopherson T (1987) The tibial subchondral plate. J Bone Joint Surg Am 69:1212–1220

Eckstein F, Müller-Gerbl M, Putz R (1992) Distribution of subchondral bone density and cartilage thickness in the human patella. J Anat 180 (Pt 3):425–433

119

Eckstein F, Lohe F, Schulte E, Müller-Gerbl M, Milz S, Putz R (1993a) Physiological incongruity of the humero-ulnar joint: a functional principle of optimized stress distribution acting upon articulating surfaces? Anat Embryol (Berl) 188:449–455

Eckstein F, Lohe F, Steinlechner M, Müller-Gerbl M, Putz R (1993b) Kontaktflächen des menschlichen Humeroulnargelenks in Abhängigkeit von der Anpresskraft, ihr Zusammenhang mit subchondraler Mineralisierung und Gelenkflächenmorphologie der Incisura trochlearis. Anat Anz 175:545–552

Eckstein F, Müller-Gerbl M, Steinlechner M, Benedetto K P, Putz R (1993c) Subchondrale Mineralisierungsmuster bei Chondromalacia patellae. Arthroskopie 6:116–120

Eckstein F, Putz R, Müller-Gerbl M, Steinlechner M, Benedetto KP (1993d) Cartilage degeneration in the human patellae and its relationship to the mineralisation of the underlying bone: a key to the understanding of chondromalacia patellae and femoropatellar arthrosis? Surg Radiol Anat 15:279–286

Eckstein F, Steinlechner M, Müller-Gerbl M, Putz R (1993e) Mechanische Beanspruchung und subchondrale Mineralisierung des menschlichen Ellbogengelenks. Eine CT-osteoabsorptiometrische Studie. Unfallchirurg 96:399–404

Eckstein F, Merz B, Schmid P, Putz R (1994a) The influence of geometry on the stress distribution in joints–a finite element analysis. Anat Embryol (Berl) 189:545–552

Eckstein F, Müller-Gerbl M, Putz R (1994b) Die Verteilung der Knorpeldegeneration an der menschlichen Patella in Beziehung zur individuellen subchondralen Mineralisierung. Z Orthop Ihre Grenzgeb 132:405–411

Eckstein F, Lohe F, Hillebrand S, Bergmann M, Schulte E, Milz S, Putz R (1995a) Morphomechanics of the humero-ulnar joint: I. Joint space width and contact areas as a function of load and flexion angle. Anat Rec 243:318–326

Eckstein F, Merz B, Müller-Gerbl M, Holzknecht N, Pleier M, Putz R (1995b) Morphomechanics of the humero-ulnar joint: II. Concave incongruity determines the distribution of load and subchondral mineralization. Anat Rec 243:327–335

Eckstein F, Müller-Gerbl M, Steinlechner M, Kierse R, Putz R (1995c) Subchondral bone density in the human elbow assessed by computed tomography osteoabsorptiometry: a reflection of the loading history of the joint surfaces. J Orthop Res 13:268–278

Eckstein F, Merz B, Sittek H, Kolem H, Reiser M, Putz R (1996) Geometry-to-pressure relationship in the human elbow joint – a qualitative analysis using MRI and finite elements. Eur J Anat 1:23–30

Eckstein F, Merz B, Jacobs C, Schön M, Putz R (1997) Zugspannungen bestimmen die funktionelle Anpassung des subchondralen Knochens in inkongruenten Gelenken. Osteologie 6[Suppl]:30

Eisenhart-Rothe R van, Eckstein F, Löhe F, Landgraf J, Müller-Gerbl M, Putz R (1996) Verteilung der anatomischen Gelenkspaltweite und Flächenpressung im menschlichen Hüftgelenk – eine quantitative Analyse. Osteologie 5:55–64

Eisenhart-Rothe von R, Eckstein F, Müller-Gerbl M, Landgraf J, Rock C, Putz R (1997) Direct comparison of contact areas, contact stress and subchondral mineralization in human hip joint specimens. Anat Embryol 195:279–288

Ekholm R, Norbäck B (1951) On the relationship between articular changes and function. Acta Orthop Scand 21:81–98

Ekholm R, Ingelmark BE (1952) Functional thickness variations of human articular cartilage. Acta Soc Med Ups 57:39–59

el-Bohy AA, Yang KH, King AI (1989) Experimental verification of facet load transmission by direct measurement of facet lamina contact pressure. J Biomech 22:931–941

Evans FG (1969) The mechanical properties of bone. Artif Limbs 13:37–48

Evans FG, Bang S (1967) Differences and relationships between the physical properties and the microscopic structure of human femoral, tibial, and fibular cortical bone. Am J Anat 120:79–88

Ewald FC, Poss R, Pugh J, Schiller AL, Sledge CB (1982) Hip cartilage supported by methacrylate in canine arthroplasty. Clin Orthop 171:273–279

Fairbank TJ (1948) Knee joint changes after meniscectomy. J Bone Joint Surg Br 30:664–670

Farkas T, Boyd RD, Schaffler MB, Radin EL, Burr DB (1987) Early vascular changes in rabbit subchondral bone after repetitive impulsive loading. Clin Orthop 219:259–267

Favenesi JA, Gardeniers JWM, Huiskes R, Slooff TJJH (1984) Mechanical properties of normal and avascular cancellous bone. Trans Orthop Res Soc 8:198

Fick R (1911) Spezielle Gelenk- und Muskelmechanik In: Handbuch der Anatomie des Menschen Bd II, Abt. 1, 3. , Fischer, Jena

Firoozbakhsh K, Cowin SC (1981) An analytical model of Pauwel's functional adaptation mechanism in bone. J Biomech Eng 103:246–252

Frahm R (1991) Experimentelle und klinische Studien zur Rotationsfehlstellung nach distaler Radiusfraktur (Habilitationsschrift), Albert-Ludwig-Universität, Freiburg

Frank EH, Grodzinsky AJ (1987a) Cartilage electromechanics–I. Electrokinetic transduction and the. J Biomech 20:615–627

Frank EH, Grodzinsky AJ (1987b) Cartilage electromechanics–II. A continuum model of cartilage. J Biomech 20:629–639

Frercks J (1966) Vergleichende chemisch-analytische Untersuchungen des spongiösen und kompakten Knochens aus fünf verschiedenen Skelettabschnitten. Inaug.Diss, Christian-Albrechts-Universität, Kiel

Frost HM (1986) Intermediary organization of the skeleton. CRC Press, Boca Raton, Florida

Fukubayashi T, Kurosawa H (1980) The contact area and pressure distribution pattern of the knee. A study of normal and osteoarthrotic knee joints. Acta Orthop Scand 51:871–879

Fyhrie DP, Carter DR (1986) A unifying principle relating stress to trabecular morphology. J Orthop Res 4:304–317

Galante J, Rostoker W, Ray RD (1970) Physical properties of trabecular bone. Calcif Tissue Res 5:236–246

Gardner DL (1965) Pathology of the connective tissue diseases. Arnold, London

Genant HK, Boyd D (1977) Quantitative bone mineral analysis using dual energy computed. Invest Radiol 12:545–551

Giunta R, Löwer N, Kierse R, Wilhelm K, Müller-Gerbl M (1997a) Die Beanspruchung des Radiokarpalgelenks-CT-Untersuchungen der subchondralen Knochendichte in vivo. Handchir Mikrochir Plast Chir 29:32–37

Giunta R, Löwer N, Wilhelm K, Kierse R, Rock C, Müller-Gerbl M (1997b) Altered patterns of subchondral bone mineralization in kienböck's disease. J Hand Surg 22-B:16–20

Gluer CC, Reiser UJ, Davis CA, Rutt BK, Genant HK (1988) Vertebral mineral determination by quantitative computed tomography (QCT): accuracy of single and dual energy measurements. J Comput Assist Tomogr 12:242–258

Goel VK, Singh D, Bijlani V (1982) Contact areas in human elbow joints. J Biomech Eng 104:169–175

Goldstein SA, Wilson DL, Sonstegard DA, Matthews LS (1983) The mechanical properties of human tibial trabecular bone as a function of metaphyseal location. J Biomech 16:965–969

Gong JK, Arnold JS, Cohn SH (1964) Composition of trabecular and cortical bone. Anat Rec 149:325–332

Goodfellow JW, Bullough PG (1967) The pattern of ageing of the articular cartilage of the elbow joint. J Bone Joint Surg Br 49:175–181

Goodfellow JW, Hungerford D, Zindel M (1971) Chondromalacia Patellae. Orthopaedics: Oxford Volume 4:111

Goodship AE, Lanyon LE, MacFie JH (1979) Functional adaptation of bone to increased stress. J Bone Joint Surg Am 61:539–546

Green WT Jr, Martin GN, Eanes ED, Sokoloff L (1970) Microradiographic study of the calcified layer of articular. Arch Pathol 90:151–158

Greenwald AS, Haynes DW (1972) Weight-bearing areas in the human hip joint. J Bone Joint Surg Br 54:157–163

Grüber J, Wolter D, Lierse W (1986) Der vordere Kreuzbandreflex (LCA-Reflex). Unfallchirurg 89:551–554

Grynpas MD, Alpert B, Katz I, Lieberman I, Pritzker KP (1991) Subchondral bone in osteoarthritis. Calcif Tissue Int 49:20–26

Hadler NM, Gillings DB, Imbus HR, Levitin PM, Makuc D, Utsinger PD (1978) Hand structure and function in an industrial setting. Arthritis Rheum 21:210–220

Hara T, Horii E, An KN, Cooney WP, Linscheid RL, Chao EY (1992) Force distribution across wrist joint: application of pressure-sensitive conductive rubber. J Hand Surg Am 17:339–347

Harada Y, Wevers, HW, Cooke TDV (1988) Distribution of bone strength in the proximal tibia. J Arthroplasty 3:167–175

Harell A, Dekel S, Binderman I (1977) Biochemical effect of mechanical stress on cultured bone cells. Calcif Tissue Res 22 [Suppl]:202–207

Harrington IJ (1983) Static and dynamic loading patterns in knee joints with deformities. J Bone Joint Surg Am 65:247–259

Harrison MHM, Schajowicz F, Trueta J (1953) Osteoarthritis of the hip: a study of the nature and evolution of the disease. J Bone Joint Surg Br 35:598–626

Haus J, Halata Z (1990) Innervation of the anterior cruciate ligament. International Orthopaedics 14:293–296

Havdrup T, Hulth A, Telhag H (1976) The subchondral bone in osteoarthritis and rheumatoid arthritis of. Acta Orthop Scand 47:345–350

Hayes WC, Carter DR (1976) Postyield behavior of subchondral trabecular bone. J Biomed Mater Res 10:537–544

Hayes WE, Snyder BD (1981) Toward a quantitative formulation of Wolff's law in trabecular bone. In: Cowin SC (ed) Mechanical properties of bone AMD Vol.45 American Society of Mechanical Engineers, New York, pp43–68

Haynes DW, Woods CG (1975) Nutritional pathways for adult articular cartilage. Orthop Oxford 8:1–8

Hegedus DM, Cowin SC (1976) Bone remodeling I: Theory of adaptive elasticity. II: Small strain adaptive elasticity. J Elasticity 6:313–325

Hehne HJ (1983) Das Patellofemoralgelenk. Enke, Stuttgart

Heine I (1925) Über die Querfurche am Olecranon. Anat Anz 59:257–271

Hellmann DB, Helms CA, Genant HK (1983) Chronic repetitive trauma: a cause of atypical degenerative joint. Skeletal Radiol 10:236–242

Heuck F, Schmidt E (1960) Konzentration und Verteilung der Kalksalze in der Knochenmatrix bei Osteopathien. Verh Dtsch Orthop Ges 48:201–209

Holmdahl DE, Ingelmark BE (1948) Der Bau des Gelenkknorpels unter verschiedenen funktionellen Verhältnissen. Experimentelle Untersuchungen an wachsenden Kaninchen. Acta Anat 6:309–375

Holmdahl DE, Ingelmark BE (1950) The contact between the articular cartilage and the medullary cavities of bone. Acta Orthop Scand 20:156–165

Horii E, Garcia-Elias M, Bishop AT, Cooney WP, Linscheid RL, Chao EY (1990) Effect on force transmission across the carpus in procedures used to treat Kienböck's disease. J Hand Surg Am 15:393–400

Hübener KH (1981) Computertomographie des Körperstammes. In: W Frommhold (ed) Röntgen wie? wann? Bd VI, Thieme, Stuttgart

Hulth A (1993) Does osteoarthrosis depend on growth of the mineralized layer of cartilage? Clin Orthop 287:19–24

Hunter D, MCLaughlin AIG, Perry KMA (1945) Clinical effects of the use of pneumatic tools. Br J Indust Med 2:10–16

Hvid I (1988) Mechanical strength of trabecular bone at the knee. Laegeforeningens Forlag, Arhus

Hvid I, Hansen SL (1985) Trabecular bone strength patterns at the proximal tibial epiphysis. J Orthop Res 3:464–472

Hvid I, Hansen SL (1986) Subchondral bone strength in arthrosis. Cadaver studies of tibial condyles. Acta Orthop Scand 57:47–51

Hvid I, Bentzen SM, Jorgensen J (1986) Remodeling of trabecular bone at the proximal tibia after total knee replacement. A CT-scan study. Eng Med 15:89–93

Hvid I, Bentzen SM, Jorgensen J (1988) Remodeling of the tibial plateau after knee replacement. CT bone densitometry. Acta Orthop Scand 59:567–573

Hvid I, Bentzen SM, Linde F, Mosekilde L, Pongsoipetch B (1989) X-ray quantitative computed tomography: the relations to physical. J Biomech 22:837–844

Inaba H (1996) Distributions of subchondral bone strength and cartilage thickness of the knee joint. 10th Conference of the Eur Society of Biomechanics, pp181

Ingelmark BE (1950) The nutritive supply and nutritional value of synovial fluid. Acta Orthop Scand 20:144–155

Ingelmark BE, Ekholm R (1948) A study on variations in the thickness of articular cartilage in association with rest periodical loading. An experimental investigation on rabbits. Acta Soc Med Ups 53:61–74

Inoue H (1981) Alterations in the collagen framework of osteoarthritic cartilage and subchondral bone. Int Orthop 5:47-52

Isherwood I, Rutherford RA, Pullan BR, Adams PH (1976) Bone-mineral estimation by computer-assisted transverse axial. Lancet 2:712-715

Issekutz B, Blizzard J, Birkhead N, Rodahl K (1966) Effect of prolonged bed rest on urinary calcium output. J Appl Physiol 21:1013-1020

Jacob HA, Huggler AH, Dietschi C, Schreiber A (1976) Mechanical function of subchondral bone as experimentally determined on the acetabulum of the human pelvis. J Biomech 9:625-627

Jacobs CR, Eckstein F (1997) Computer simulation of subchondral bone adaption to mechanical loading in an incongruous joint. Anat Rec (in press)

Jaskulka RA, Ittner G, Schedl R (1989) Fractures of the posterior tibial margin: their role in the prognosis of malleolar fractures. J Trauma 29:1565-1570

Jasty M, Harrigan TP, Greer JA, Chen J, Harris WH (1985b) Prediction of directional variations in material properties of human cancellous bone using 3-D stereologic technique. Trans Orthop Res Soc 9:352

Jasty M, Harrigan TP, Mann RW, Harris WH (1985a) Characterization of microstructural anisotropy in cancellous bone using a second rank tensor. Trans Orthop Res Soc 9:375

Johns HE, Cunningham, JR (1983) The physics of radiology, 4th en, Charles Thomas, Illinois

Johnson F, Leitl S, Waugh W (1980) The distribution of load across the knee. A comparison of static and dynamic measurements. J Bone Joint Surg Br 62:346-349

Jones H, Priest J, Hayes W, Tichenor C, Nagel D (1977) Humeral hypertrophy in response to exercise. J Bone Joint Surg Am 59:204-208

Kalender WA, Suess C (1987) A new calibration phantom for quantitative computed tomography. Med Phys 14:863-866

Kalender WA, Perman WH, Vetter JR, Klotz E (1986) Evaluation of a prototype dual-energy computed tomographic apparatus. I.Phantom studies. Med Phys 13:334-339

Kamibayashi L, Wyss UP, Cooke TD, Zee B (1995) Changes in mean trabecular orientation in the medial condyle of the proximal tibia in osteoarthritis. Calcif Tissue Int 57:69-73

Kaplan SJ, Hayes WC, Stone JL, Beaupre GS (1985) Tensile strength of bovine trabecular bone. J Biomech 18:723-727

Kellam JF, Waddell JP (1979) Fractures of the distal tibial metaphysis with intra-articular extension-the distal tibial explosion fracture. J Trauma 19:593-601

Kern D, Zlatkin MB, Dalinka MK (1988) Occupational and post-traumatic arthritis. Radiol Clin North Am 26:1349-1358

Kettelkamp DB, Chao EY (1972) A method for quantitative analysis of medial and lateral compression forces at the knee during standing. Clin Orthop 83:202-213

Klawitter JJ, Weinstein AM (1975) Structure and properties of subchondral cancellous bone . Am J Surg 57:578

Knief J-J (1967a) Quantitative Untersuchung der Verteilung der Hartsubstanzen im Knochen und ihrer Beziehung zur lokalen mechanischen Beanspruchung. Z Anat Entwickl-Gesch 126:55-80

Knief J-J (1967b) Materialverteilung und Beanspruchungsverteilung im coxalen Femurende – Densitometrische und spannungsoptische Untersuchungen. Z Anat Entwickl-Gesch 126:81-116

Koch JC (1917) The laws of bone architecture. Am J Anat 21:177

Koebke J, Fehrmann P, Mockenhaupt J (1989) Zur Beanspruchung des normalen und des pathologischen Handgelenks. Handchir Mikrochir Plast Chir 21:127-133

Kohn D, Glaubitz W, Schmidt H, Lobenhofer P (1985) Lokalisation, Art und endoskopische Beurteilung degenerativer Veränderungen im Schultergelenk. In: HJ Refior, W Plitz, M Jäger, MH Hackenbroch (eds), Biomechanik der gesunden und kranken Schulter. Thieme, Stuttgart, pp152-156

Konermann H (1970) Dichteverteilung im Röntgenbild des Skeletts. Naturwissenschaften. 57:255

Konermann H (1971) Quantitative Bestimmung der Materialverteilung nach Röntgenbildern des Knochens mit einer neuen photographischen Methode. Z Anat Entwickl-Gesch 134:13-48

Korenstein R, Somjen D, Laub F, Danon A, Fischler H, Bindermann I (1983) Pulsed external electric fields are mitogens for bone cells. In: Oplatka A, Balaban AT (eds) Biological structures and coupled flows, Academic Press, New York, pp401-411

Kouris K, Spyrou NM, Jackson DF (1982) Imaging with ionizing radiations. Surrey University press. Glasgow, London. 1st edn

Krauspe R, Schmidt M, Schaible HG (1993) Tierexperimentelle elektrophysiologische Charakterisierung sensorischer Afferenzen aus dem vorderen Kreuzband. Orthopädie Mitteilungen 2:103–104

Kufahl RH, Saha S (1990) A theoretical model for stress-generated fluid flow in the canaliculi-lacunae network in bone tissue. J Biomech 23:171–180

Kummer B (1962) Funktioneller Bau und funktionelle Anpassung des Knochens. Anat Anz 111:261–293

Kummer B (1968) Die Beanspruchung des menschlichen Hüftgelenks. I.Allgemeine Problematik. Z Anat Entwickl-Gesch 127:277–285

Kummer B (1972) Biomechanics of bone: mechanical properties, functional structure, functional adaptation. In: Fung YC,Perrone N,Anliker M.(eds) Biomechanics: its foundations and objectives. Prentice Hall, Englewood Cliffs, pp237–271

Kummer B (1978) Mechanische Beanspruchung und funktionelle Anpassung des Knochens. Verh Anat Ges 72:21–46

Kummer B (1981) Biomechanik der Wirbelgelenke. In: Junghanns H (ed) Die Wirbelsäule in Forschung und Praxis, Bd 87, Hippokrates, Stuttgart, pp29–34

Kummer B (1985) Kausale Histogenese der Gewebe des Bewegungsapparates und funktionelle Anpassung. In: Staubesand J (ed) Benninghoff,Anatomie Bd. 1. Urban and Schwarzenberg, München Wien Baltimore, pp 199–213

Kummer B (1987) Anatomie und Biomechanik des Kniegelenksmeniscus. Langenbecks Arch Chir 372:241–246

Kummer B (1992) Biomechanische Probleme der aufrechten Haltung. Vortrag bei der 86. Versammlung der Anatomischen Gesellschaft Szeged. Ann Anat 174:33–39

Kurrat HJ (1977) Die Beanspruchung des menschlichen Hüftgelenkes. VI. Eine funktionelle Analyse der Knorpeldickenverteilung am menschlichen Femurkopf. Anat Embryol 150:129–140

Kurrat HJ, Oberländer W (1978) The thickness of the cartilage in the hip joint. J Anat 126:145–155

Lane LB, Bullough PG (1980) Age-related changes in the thickness of the calcified zone and the number of tidemarks in adult human articular cartilage. J Bone Joint Surg Br 62:372–375

Lane LB, Villacin A, Bullough PG (1977) The vascularity and remodelling of subchondral bone and calcified cartilage in adult human femoral and humeral heads. An age- and stress-related phenomenon. J Bone Joint Surg Br 59:272–278

Lanyon LE, Goodship AE, Pye CJ, MacFie JH (1982) Mechanically adaptive bone remodelling. J Biomech 15:141–145

Layton MW, Goldstein SA, Goulet RW, Feldkamp LA, Kubinski DJ, Bole GG (1988) Examination of subchondral bone architecture in experimental osteoarthritis by microscopic computed axial tomography. Arthritis Rheum 31:1400–1405

Lee JK, Yao L (1988) Stress fractures: MR imaging. Radiology 169:217–220

Lehmann J (1990) Untersuchungen zur Beanspruchung des intakten und in Fehlstellung verheilten distalen Radius. Dissertation Köln

Lehmann LA, Alvarez RE, Macovski A, Brody WR, Pelc NJ, Riederer SJ, Hall AL (1981) Generalized image combinations in dual-kVp digital radiography. Med Phys 8:659–667

Lereim P, Goldie I, Dahlberg E (1974) Hardness of the subchondral bone of the tibial condyles in the normal state and in osteoarthritis and rheumatoid arthritis. Acta Orthop Scand 45:614–627

Lereim P, Goldie IF (1975) Relationship between morphologic features and hardness of the. Arch Orthop Unfallchir 81:1–11

Lin HS, Liu YK, Adams KH (1978) Mechanical response of the lumbar intervertebral joint under physiological (complex) loading. J Bone Joint Surg Am 60:41–55

Linde F, Hvid I, Jensen NC (1986) Mechanical properties of trabecular bone in repetitive axial loading. In: Christl P, Meunier A, Lee AJC (eds) Biological and Biomechanical Performance of Biomaterial. Elsevier, Amsterdam, pp 447

Lotz JC, Gerhart TN, Hayes WC (1991) Mechanical properties of metaphyseal bone in the proximal femur. J Biomech 24:317–329

Lutz G (1967) Die Entwicklung der kleinen Wirbelgelenke. Z Orthop Ihre Grenzgeb 104:19–28

Mankin HJ (1974) The reaction of articular cartilage to injury and osteoarthritis. N Engl J Med 291:1285–1292

Maquet P (1976) Biomechanics of the knee. Springer, Berlin-Heidelberg-New York. French Edition (1977): Biomécanique du genou, Springer

Maquet GJ (1984) Biomechanics of the knee. 2nd Springer Verlag, Berlin

Maquet PG, Van de Berg AJ, Simonet JC (1975) Femorotibial weight-bearing areas. J Bone Joint Surg Am 57:766–771

Martin R, Albright J, Clarke W, Niffenegger J (1981) Load-carrying effects on the adult beagle tibia. Sports Exer 13:343–349

Martin RB, Ishida J (1989) The relative effects of collagen fiber orientation, porosity, density, and mineralization on bone strength. J Biomech 22:419–426

Mayor MB, Moskowitz RW (1974) Metabolic studies in experimentally-induced degenerative joint disease in the rabbit. J Rheumatol 1:17–23

McCullough EC (1982) X-ray transmission CT scanner survey. J Comput Assist Tomogr 6:423–428

Meachim G, Allibone R (1984) Topographical variation in the calcified zone of upper femoral articular cartilage. J Anat 139:341–352

Meema HE, Harris CK, Porett RE (1964) A method for determination of bone-salt content of cortical bone. Radiology 82:986–997

Mente PL, Lewis JL (1994) Elastic modulus of calcified cartilage is an order of magnitude less than that of subchondral bone. J Orthop Res 12:637–647

Merz B, Eckstein F, Hillebrand S (1997) Mechanical implications of humero-ulnar incongruity – finite element analysis and experiment. J Biomechanics 30:713–721

Mezaros T, Vizkelety T (1986) Structure of the subchondral bone plate, Xvth. Symposium of the European Society of Osteoarthrology, Publications of the University of Kuopio Medicine

Milz S, Putz R (1994a) Quantitative morphology of the subchondral plate of the tibial plateau. J Anat 185:103–110

Milz S, Putz R (1994b) Lückenbildungen der subchondralen Mineralisierungszone des Tibiaplateaus. Osteologie 3:110–118

Milz S, Eckstein F, Putz R (1995) The thickness of the subchondral plate and its correlation with the thickness of the uncalcified articular cartilage in the human patella. Anat Embryol Berl 192:437–444

Milz S, Eckstein F, Putz R (1997) Thickness distribution of the subchondral mineralization zone of the trochlear notch and ist correlation with the cartilage thickness: An expression of functional adaptation to mechanical stress acting on the humeroulnar joint? Anat Rec 248:1–9

Mink JH, Deutsch AL (1989) Occult cartilage and bone injuries of the knee: detection, classification, and assessment with MR imaging. Radiology 170:823–829

Mital MA, Millington PF (1971) Surface characteristics of articular cartilage. Micron 2:236–249

Miyanaga Y, Fukubayashi T, Kurosawa H (1984) Contact studies of the hip joint. Arch Orthop Trauma Surg 103:13–17

Mohr W (1984) Gelenkerkrankungen. Thieme, Stuttgart – New York

Moller BN, Krebs B (1982) Intra-articular fractures of the distal tibia. Acta Orthop Scand 53:991–996

Möllers N, Lehmann K, Koebke J (1986) Die Verteilung des subchondralen Knochenmaterials an der distalen Gelenkfläche des Radius. Anat Anz 161:151

Molzberger H (1973) Die Beanspruchung des menschlichen Hüftgelenks. IV.Analyse der funktionel-len Struktur der Tangentialfaserschicht des Hüftpfannenknorpels. Z Anat Entwickl-Gesch 139:283–240

Müller W (1982) Das Knie. Springer, Heidelberg

Müller-Gerbl M (1992) Die Rolle der Wirbelgelenke für die Kinematik der Bewegungssegmente. Ann Anat 174:48–53

Müller-Gerbl M, Schulte E, Putz R (1987a) The thickness of the calcified layer of articular cartilage: a function of the load supported ? J Anat 154:103–111

Müller-Gerbl M, Schulte E, Putz R (1987b) The thickness of the calcified layer in different joints of a single individual. Acta Morphol Neerl Scand 25:41–49

Müller-Gerbl M, Putz R, Hodapp N, Schulte E, Wimmer B (1989) Computed tomography-osteoab-sorptiometry for assessing the density distribution of subchondral bone as a measure of long-term mechanical adaptation in individual joints. Skeletal Radiol 18:507–512

Müller-Gerbl M, Putz R, Hodapp N, Schulte E, Wimmer B (1990a) Computed tomography-osteoabsorptiometry: a method of assessing the mechanical condition of the major joints in a living subject. Clinical Biomech 5:193–198

Müller-Gerbl M, Putz R, Hodapp N, Schulte E, Wimmer B (1990b) Die Darstellung der subchondralen Dichtemuster mittels der CT-Osteoabsorptiometrie (CT OAM) zur Beurteilung der individuellen Gelenkbeanspruchung am Lebenden. Z Orthop 128:128–133

Müller-Gerbl M, Putz R, Boscher HP, Schweizer L (1990c) Beanspruchung des Schulter- und des AC-Gelenkes bei Spitzenturnern. In: Bernett P, Jeschke D (eds) Sport und Medizin, Pro und Contra, Zuckerschwerdt Verlag, München, pp256–258

Müller-Gerbl M, Putz R, Schulte E (1990d) Die Verteilungsmuster des Knorpels und der subchondralen Knochendichte als morphologische Parameter der individuellen Gelenkbeanspruchung. In: W Glinz (ed) Fortschritte in der Arthroskopie, 6:8–13

Müller-Gerbl M, Hodapp N, Reinbold WD, Putz R (1991) Can CT osteoabsorptiometry be used to display the distribution of subchondral mineralisation ? Calcif Tissue Int 48[Suppl]:68

Müller-Gerbl M, Putz R, Kenn R (1992) Demonstration of subchondral bone density patterns by three-dimensional CT osteoabsorptiometry as a noninvasive method for in vivo assessment of individual long-term stresses in joints. J Bone Miner Res 7 [Suppl 2]:411–418

Müller-Gerbl M, Putz R, Kenn R (1993a) Verteilungsmuster der subchondralen Mineralisierung in der Cavitas Glenoidalis bei Normalpersonen, Sportlern und Patienten. Z Orthop Ihre Grenzgeb 131:10–13

Müller-Gerbl M, Putz R, Kenn R, Kierse R (1993b) People in different age groups show different hip-joint morphology. Clin Biomech 8:66–72

Mullender MG, Huiskes R, Weinans H (1994) A physiological approach to the simulation of bone remodeling as a self-organizational control process. J Biomechanics 27:1389–1394

Murray RP, Hayes WC, Edwards WT, Harry JD (1984) Mechanical properties of the subchondral plate and the metaphyseal shell. Trans 30th Annual Orthop Res Soc 9:197

Nilsson B, Westlin N (1971) Bone density in athletes. Clin Orthop 77:179–182

Noble J, Alexander K (1985) Studies of tibial subchondral bone density and its significance. J Bone Joint Surg Am 67:295–302

Noyes FR (1977) Functional properties of knee ligaments and alterations induced by immobilization. Clin Orthop 123:210–242

Oberländer W (1973) Die Beanspruchung des menschlichen Hüftgelenks. V. Die Verteilung der Knochendichte im Acetabulum. Z Anat Entwickl-Gesch 140:367–384

Oberländer W (1977) Die Beanspruchung des menschlichen Hüftgelenks. VIII. Die Verteilung der Knorpeldicke im Acetabulum und ihre funktionelle Deutung. Anat Embryol 150:141–153

Oberländer W (1978) On Biomechanics of joints, the influence of functional swelling on the congruity of regularly curved joints. J Biomech 11:151–153

Oberländer W, Kurrat HJ (1979) Die Knorpeldickenverteilung im distalen Anteil des Ellbogengelenkes und ihre Beziehung zur Lage degenerativer Knorpelveränderungen. Verh Anat Ges 73:891–895

Odgaard A, Pedersen CM, Bentzen SM, Jorgensen J, Hvid I (1989) Density changes at the proximal tibia after menisectomy. J Orthop Res 7:744–753

Oettmeier R, Abendroth K, Oettmeier S (1989) Analyses of the tidemark on human femoral heads. II. Tidemark. Acta Morphol Hung 37:169–180

Oettmeier R, Arokuski J, Roth AJ, Helminen HJ, Tammi M, Abendroth K (1992) Quantitative study of articular cartilage and subchondral bone remodeling in the knee joint of dogs after strenuous running training. J Bone Miner Res 7:419–424

Ogata K, Whiteside LA, Lesker PA, Simmons DJ (1977) The effect of varus stress on the moving rabbit knee joint. Clin Orthop 129:313–318

Ogston A (1878) On the growth and maintenance of the articular ends of adult bones. J Anat Physiol 12:503

Outerbridge RE (1961) The etiology of chondromalacia patellae. J Bone Joint Surg Br 43:752–757

Palmer AK, Werner FW (1984) Biomechanics of the distal radioulnar joint. Clin Orthop 187:26–35

Parniapour M, Nordin M, Skovron ML, Frankel VH (1990) Environmentally induced disorders of the musculoskeletal system. Med Clin North Am 74:347–359

Paul B (1975) Elektronenmikroskopische Befunde zur Chondromalacia patellae und ihre Bedeutung für das Arthrose-Problem – Diskussion zu unterschiedlichen Altersstufen. Beitr Orthop Traumatol 22:560–565

Pauwels F (1955) Über die Verteilung der Spongiosadichte im coxalen Femurende und ihre Bedeutung für die Lehre vom funktionellen Bau des menschlichen Knochens. 7.Beitrag zur funktionellen Anatomie und kausalen Morphologie des Stützapparates. Morph Jb 95:35–54

Pauwels F (1960) Eine neue Theorie über den Einfluß mechanischer Reize auf die Differenzierung der Stützgewebe. 10. Beitrag zur funktionellen Anatomie und kausalen Morphologie des Stützapparates. Z Anat Entwickl-Gesch 121:478–515

Pauwels F (1963) Die Druckverteilung im Ellbogengelenk, nebst grundsätzlichen Bemerkungen über den Gelenkdruck. 11.Beitrag zur funktionellen Anatomie und kausalen Morphologie des Stützapparates. Z Anat Entwickl-Gesch 123:643–667

Pauwels F (1965) Gesammelte Abhandlungen zur funktionellen Anatomie des Bewegungsapparates. Springer Berlin-Heidelberg-New York

Pauwels F (1973) Atlas zur Biomechanik der gesunden und kranken Hüfte. Springer, Berlin-Heidelberg-New York

Pauwels F (1980) Biomechanics of the locomotor apparatus. Springer Berlin.

Pedersen DR, Crowninshield RD, Brand RA, Johnston RC (1982) An axisymmetric model of acetabular components in total hip arthroplasty. J Biomech 15:305–315

Pedley RB, Meachim G (1979) Topographical variation in patellar subarticular calcified tissue density. J Anat 128:737–745

Piatkowski J, Gräwe A, Ehler E, Schumacher GH (1985) Regionale Unterschiede in der chemischen Zusammensetzung der menschlichen Tibia. Anat Anz 158:315–322

Prendergast PJ, Huiskes R, Soballe K (1996) Biomechanical influences on tissue differentiation at implant interfaces. In: Transaction of the 42nd Annual Meeting, Orthopaedic Research Society 21:527

Pugh JW, Rose RM, Radin EL (1973a) A possible mechanism of Wolff's law: trabecular microfractures. Arch Int Physiol Biochim 81:27–40

Pugh JW, Rose RM, Radin EL (1973b) Elastic and viscoelastic properties of trabecular bone: dependence on structure. J Biomech 6:475–485

Pugh JW, Radin EL, Rose RM (1974) Quantitative studies of human subchondral cancellous bone. Its relationship to the state of its overlying cartilage. J Bone Joint Surg Am 56:313–321

Putti V (1937) Die Anatomie der angeborenen Hüftverrenkung. Enke, Stuttgart

Putz R (1976) Zur Morphologie und Rotationsmechanik der kleinen Gelenke der Lendenwirbel Z Orthop 114:902–912

Putz R (1981) Funktionelle Anatomie der Wirbelgelenke, In: Doerr W, Leonhardt H (eds) Normale und Pathologische Anatomie, Vol 43, Thieme Stuttgart

Putz R (1985) The functional morphology of the superior articular processes of the lumbar vertebrae. J Anat 143:181–187

Putz R, Müller-Gerbl M, Schulte E, Wimmer B (1987) Verteilung der Knorpeldicke und der Mineralisierung im Kniegelenk. In: HJ Refior, MH Hackenbroch, CJ Wirth (eds). Der alloplastische Ersatz des Kniegelenkes... Thieme, Stuttgart – New York, pp2–5

Putz R, Eckstein F, Müller-Gerbl M, Benedetto K (1990) The subchondral bone density and arthroscopic findings in femoro-patellar joint.J Biomech 24:487

Radin EL (1972) Role of mechanical factors in the pathogenesis of primary osteoarthrosis. Lancet 1:519–522

Radin EL, Paul IL (1970) Does cartilage compliance reduce skeletalimpact loads ? The relative force-attenuating properties of articular cartilage, synovial fluid, periarticular soft-tissues and bone. Arthritis Rheum 13:139–144

Radin EL, Rose RM (1986) Role of subchondral bone in the initiation and progression of cartilage damage. Clin Orthop 213:34–40

Radin EL, Paul IL, Tolkoff MJ (1970) Subchondral bone changes in patients with early degenerative joint disease. Arthritis Rheum 13:400–405

Radin EL, Parker HG, Pugh JW, Steinberg RS, Paul IL, Rose RM (1973) Response of joints to impact loading. 3. Relationship between trabecular microfractures and cartilage degeneration. J Biomech 6:51–57

Radin EL, Paul IL, Rose RM (1975) Mechanical factors in the aetiology of osteoarthrosis. Ann Rheum Dis 34[Suppl]:132–133

Radin EL, Martin RB, Burr DB, Caterson B, Boyd RD, Goodwin C (1984) Effects of mechanical loading on the tissues of the rabbit knee. J Orthop Res 2:221–234

Ravelli A (1955) Das Vakuum-Phänomen (R. Ficksches Zeichen). Fortschr Röntgenstr 83:236–240

Reich NE, Seidelmann FE, Tubbs RR, Mac-Intyre WJ, Meaney TF (1976) Determination of bone mineral content using CT scanning. Am J Roentgenol 127:593–594

Reichmann S, Berglund E, Lundgren K (1972) Das Bewegungszentrum in der Lendenwirbelsaule bei Flexion und Extension. Z Anat Entwicklungsgesch 138:283–287

Reimann I (1973) Experimental osteoarthritis of the knee in rabbits induced by alteration of the load-bearing. Acta Orthop Scand 44:496–504

Reimann I, Christensen SB (1979) A histochemical study of alkaline and acid phosphatase activity in subchondral bone from osteoarthrotic human hips. Clin Orthop 140:85–91

Reimann I, Mankin HJ, Trahan C (1977) Quantitative histologic analyses of articular cartilage and subchondral bone from osteoarthritic and normal human hips. Acta Orthop Scand 48:63–73

Riede UN, Heitz P, Ruedi T (1971) Gelenkmechanische Untersuchungen zum Problem der posttrau-matischen Arthrosen im oberen Sprunggelenk. II. Einfluss der Talusform auf die Biomechanik des oberen Sprunggelenkes. Langenbecks Arch Chir 330:174–184

Robinson RA, Elliott SR (1957) The water content of bone. J Bone Joint Surg Am 39:167

Rodan GA (1981) Mechanical and electrical effects on bone and cartilage cells: translation of the physical signal into a biological message. In: Barrer HG (ed) Orthodontics, The State of The Art, The University of Pennsylvania Press, Philadelphia, pp315–322

Roux W (1912) Anpassungslehre, Histomechanik und Histochemie. Mit Bemerkungen über die Entwicklung und Formgestaltung der Gelenke. Berichtigung zu R Thomas gleichnamigem Aufsatz. Virchows Arch Path Anat 209:168–209

Rubin CT, Lanyon LE (1982) Limb mechanics as a function of speed and gait: a study of. J Exp Biol 101:187–211

Rubin CT, Lanyon LE (1987) Osteoregulatory nature of mechanical stimuli: function as a determinant for adaptive remodeling in bone. J Orthop Res 5:300–310

Ruegsegger P, Elsasser U, Anliker M, Gnehm H, Kind H, Prader A (1976) Quantification of bone mineralization using computed tomography. Radiology 121:93–97

Saha AK (1973) Mechanics of elevation of glenohumeral joint. Acta Orthop Scand 44:668–678

Salter RB, Field P (1960) The effects of continuous compression on the pressure distribution on living articular cartilage. J Bone Joint Surg Am 42:31–49

Schelble E (1987) Zur funktionellen Struktur der Wirbelbogengelenke. Dissertation Freiburg

Schleicher A, Tillmann B, Zilles K (1980) Quantitative analysis of x-ray images with a television image analyser. Microscopia Acta 83:189–196

Schmitt GH, Hübener KH (1980) Computertomographische Densitometrie formalinfixierter und gefrorener menschlicher Gewebe. Fortschr Röntgenstr. 33:531–534

Schmitt HP (1968) Über die Beziehung zwischen Dichte und Festigkeit des Knochens am Beispiel des menschlichen Femur. Z Anat Entwickl-Gesch 127:1–24

Schoenfeld CM, Lautenschlager EP, Meyer PR Jr (1974) Mechanical properties of human cancellous bone in the femoral. Med Biol Eng 12:313–317

Schünke M, Tillmann B, Schleicher A, Pointner H (1987) Biomechanische und histochemische Untersuchungen am Tibiaplateau des Menschen. Verh Anat Ges 81:451–453

Sennwald G (1987) Das Handgelenk. Springer, Berlin Heidelberg New York

Serink MT, Nachemson A, Hansson G (1977) The effect of impact loading on rabbit knee joints. Acta Orthop Scand 48:250–262

Shimizu M, Tsuji H, Matsui H, Katoh Y, Sano A (1993) Morphometric analysis of subchondral bone of the tibial condyle in osteoarthrosis. Clin Orthop 293:229–239

Shimmins J, Gillespie FC, Hamilton MD, Smith DA (1968) The measurement of bone mineral in vivo by photon absorption. Calcif Tissue Res 2[Suppl]:40

Simkin PA, Graney DO, Fiechtner JJ (1980) Roman arches, human joints, and disease: differences between convex and concave sides of joints. Arthritis Rheum 23:1308–1311

Simkin PA, Heston TF, Downey DJ, Benedict RS, Choi HS (1991) Subchondral architecture in bones of the canine shoulder. J Anat 175:213–227

Simon SR, Radin EL, Paul IL, Rose RM (1972) The response of joints to impact loading. II. In vivo behavior of subchondral bone. J Biomech 5:267–272

Singh I (1978) The architecture of cancellous bone. J Anat 127:305–310

Skerry TM, Bitensky L, Chayen J, Lanyon LE (1987) Strain memory in bone tissue ? Is proteoglycan based persistence of strain history a cue for the control of adaptive bone remodelling ? Trans Orthop Res Soc 11:277

Skerry TM, Bitensky L, Chayen J, Lanyon LE (1988) Loading related reorientation of bone proteoglycan in vivo. A strain memory in bone tissue? J Orthop Res 6:547–551

Sokoloff L (1969) The biology of degenerative joint disease. University of Chicago Press, Chicago

Sokoloff L (1974) The general pathology of osteoarthritis . In: Ali SY, Elves MW, Leaback DH (eds) Proceedings of the Symposium on Normal and Osteoarthrotic Articular Cartilage. Institute Orthopaedics, London, pp 111–123

Solomonow M, Baratta R, Zhou BH, Shoji H, Bose W, Beck C, D'Ambrosia R (1987) The synergistic action of the anterior cruciate ligament and thigh muscles in maintaining joint stability. Am J Sports Med. 15:207–213

Somjen D, Bindermann I, Berger E, Harell A (1980) Bone remodeling induced by physical stress is prostaglandin E_2 mediated. Biochim Biophys Acta 627:91–100

Soslowsky LJ, Ateshian GA, Bigliani LU, Flatow EL, Mow VC (1989) Sphericity of glenohumeral joint articulating surfaces. Transaction of the 35th Annual Meeting of the Orthopaedic Research Society, Las Vegas:228

Stafford SA, Rosenthal DI, Gebhardt MC, Brady TJ, Scott JA (1986) MRI in stress fracture. Am J Roentgenol 147:553–556

Steinberg ME, Gusenkell GL, Black J, Korostoff E (1974) Stress-induced potentials in moist bone in vitro. J Bone Joint Surg Am 65:704–713

Stevens J (1970) Osteoarthritis of the hip. A review with special consideration of. Clin Orthop 71:152–181

Stockwell RA (1987) Structure and function of the chondrocyte under mechanical stress. In: Helminen HJ, Kiviranta I, Tammi M, Säämänen A-M, Paukkonen K, Jurvelin J (eds) Joint loading:, Wright, Bristol, pp126–148

Stone JL, Beaupre GS, Hayes WC (1983) Multiaxial strength characteristics of trabecular bone. J Biomech 16:743–752

Stougard J (1974) The calcified cartilage and the subchondral bone under normal and abnormal conditions. Acta Path Microbiol Scand Section A 82:182–188

Strüter HD, Rassow J (1969) Über ein Verfahren zur quantitativen Bestimmung des Mineralsalzgehaltes der Knochen mit radioaktiven Isotopen. Fortschr Röntgenstr 110:499–506

Thompson D'AW (1942) On growth and form. Cambridge University Press, Cambridge

Thompson RC Jr, Oegema TR Jr, Lewis JL, Wallace L (1991) Osteoarthrotic changes after acute transarticular load. An animal model. J Bone Joint Surg Am 73:990–1001

Tillmann B (1969) Die Beanspruchung des menschlichen Hüftgelenks. III. Die Form der Facies lunata. Z Anat Entwickl-Gesch 128:329–349

Tillmann B (1971) Die Beanspruchung des menschlichen Ellbogengelenks. I. Funktionelle Morphologie der Gelenkflächen. Z Anat Entwickl-Gesch 134:328–342

Tillmann B (1978a) A contribution to the functional morphology of articular surfaces. Normale und Pathologische Anatomie Bd.34; W Bargmann und W Doerr (eds).; Thieme, Stuttgart

Tillmann B (1978b) Entwicklung und funktionelle Anatomie des Ellbogengelenkes. Z Orthop 116:392–400

Tillmann B (1980) Morphologische und biomechanische Untersuchungen an der Facies articularis patellae. Orthop Praxis 6:462–467

Tillmann B, Brade H (1980) Morphologische und biomechanische Untersuchungen an der Facies articularis patellae. Orthop Praxis 6:462–467

Tipton CM, James SL, Mergner W, Tscheng TK (1970) Influence of exercise on strength of medial collateral knee ligaments of dogs. Am J Physiol 218:894–901

Trueta J (1968) Studies of the development and decay of the human frame. William Heinemann Medical Books Ltd, London

Trueta J, Harrison MHM (1953) THe normal vascular anatomy of the femoral head in adult man. J Bone Joint Surg Br 35:442–461

Trumble T, Glisson RR, Seaber AV, Urbaniak JR (1987) Forearm force transmission after surgical treatment of distal radioulnar joint disorders. J Hand Surg Am 12:196–202

Tsuge S, Nakamura R (1993) Anatomical risk factors for Kienböck's disease. J Hand Surg Br 18:70–75

Ueno K, Liu YK (1987) A three-dimensional nonlinear finite element model of lumbar intervertebral joint in torsion. J Biomech Eng 109:200–209

Vellet AD, Marks PH, Fowler PJ, Munro TG (1991) Occult posttraumatic osteochondral lesions of the knee: prevalence, classification, and short-term sequelae evaluated with MR imaging. Radiology 178:271–276

Vener MJ, Thompson RC Jr, Lewis JL, Oegema TR Jr (1992) Subchondral damage after acute transarticular loading: an in vitro. J Orthop Res 10:759–765

Vetter JR, Perman WH, Kalender WA, Mazess RB, Holden JE (1986) Evaluation of a prototype dual-energy computed tomographic apparatus. II. Determination of vertebral bone mineral content. Med Phys 13:340–343

Walker C, Carpenter RJ, Oegema TR, Thompson RC (1990) Evidence for activity in the tidemark in normal articular cartilage. Transaction of the 36th Annual Meeting of the Orthopaedic Research Society, New Orleans, Louisiana

Walker PS (1973) A comparison of normal and artificial human joints. Acta Orthop Belg 39 [Suppl]:43–54

Walker PS (1977) Human joints and their artificial replacements. Thomas, Springfield

Walker PS, Hajek JV (1972) The load-bearing area in the knee joint. J Biomech 5:581–589

Walker PS, Erkman MJ (1975) The role of the menisci in force transmission across the knee. Clin Orthop 109:184–192

Wang CJ, Walker PS (1974) Rotatory laxity of the human knee joint. J Bone Joint Surg Am 56:161–170

Weaver JK (1966) The microscopic hardness of bone. J Bone Joint Surg Am 48:273–288

Werner H (1897) Die Dicke des menschlichen Gelenkknorpels. Inaug. Dissertation. Berlin

Whedon D (1984) Disuse osteoporosis: physiologic aspects. Calcif Tissue Int 36:146–150

White DR (1978) Tissue substitutes in experimental radiation physics. Med Phys 5:467–479

Wolff J (1892) Das Gesetz der Transformation der Knochen. Hirschwald, Berlin

Woo S, Kuei S, Amiel D, Gomez M, Hayes W, White F, Akeson W (1981) The effect of prolonged physical training on the properties of long bone: a study of Wolff's law. J Bone Joint Surg Am 63:780–787

Woods CG, Greenwald AS, Haynes DW (1970) Subchondral vascularity in the human femoral head. Ann Rheum Dis 29:138–142

Wu DD, Burr DB, Boyd RD, Radin EL (1990) Bone and cartilage changes following experimental varus or valgus tibial angulation. J Orthop Res 8:572–585

Wynarsky GT, Greenwald AS (1983) Mathematical model of the human ankle joint. J Biomech 16:241–252

Yang KH, Boyd RD, Kish VL, Burr DB, Caterson B, Radin EL (1989) Differential effect of load magnitude and rate on the initation and progression of osteoarthrosis. Trans Orthop Res Soc 14:148

Yates FE (1987) Control of self organization, In: Yates FE (ed) Self organizing systems, the emergence of order (Edited by Plenum Press, New York

Yao L, Lee JK (1988) Occult intraosseous fracture: detection with MR imaging. Radiology 167:749–751

Ypey DL, Weidema AF, Ravesloot JH, Nijweide PJ (1991) A role for mechanosensitive ion channels in osteoblasts and osteoclasts ? Calcif Tissue Int 48[Suppl]:89

Subject Index

Printing and binding: Druckerei Triltsch, Würzburg